Data Hiding
Fundamentals
and Applications

Content Security in Digital Media

Husrev T. Sencar
New Jersey Institute of Technology

Mahalingam Ramkumar
Mississippi State University

Ali N. Akansu
New Jersey Institute of Technology

ELSEVIER
ACADEMIC
PRESS

Amsterdam   Boston   Heidelberg   London   New York   Oxford
Paris   San Diego   San Francisco   Singapore   Sydney   Tokyo

Data Hiding Fundamentals and Applications

Elsevier Academic Press
525 B Street, Suite 1900, San Diego, California 92101-4495, USA
84 Theobald's Road, London WC1X 8RR, UK

This book is printed on acid-free paper.

Library of Congress Cataloging-in-Publication Data

Sencar, Husrev T.
    Data hiding fundamentals and applications / Husrev T. Sencar, Mahalingam Ramkumar,
Ali N. Akansu.
       p. cm.
    Includes bibliographical references and index.
    ISBN 0-12-047144-2 (alk. paper)
       1. Multimedia systems–Security measures. 2. Data encryption (Computer science) I.
Ramkumar Mahalingam. II. Akansu, Ali N., 1958- III. Title.

QA76.575.S46 2004
005.8′2–dc22
                                                                    2004052921

British Library Cataloguing in Publication Data
A catalogue record for this book is available from the British Library

ISBN: 0-12-047144-2

For all information on all Academic Press publications
visit our Web site at www.academicpress.com

Printed in the United States of America
04   05   06   07   08   9   8   7   6   5   4   3   2   1

To our wives:
Yelda
Bindu
Bilge

Contents

Type I (Linear) Data Hiding

CHAPTER 5

Type II and Type III (Nonlinear) Data Hiding Methods

CHAPTER **6**
Advanced Implementations

CHAPTER **7**
Major Design Issues

CHAPTER **8**

Data Hiding Applications

APPENDIX **A**

CAE-CID Framework under Varying Channel Noise   221

APPENDIX **B**

Statistics of $\rho_{dep}|P$
and $d_{dep}|P$ 223

APPENDIX **C**

Mathematical Proofs

Preface

Hiding anything invaluable is natural and, therefore, widely applicable in our daily lives. Data hiding (steganography) is an old concept that has been utilized since the early ages of human history in different forms and for a variety of purposes. Whenever the value of information is high, its transportation and delivery becomes an important task.

The Internet revolution, digital representation, and transportation of data offered efficient solutions for information delivery. These phenomenal developments brought their own concerns and problems in security, monitoring, and use of information by the qualified end users. Therefore, IT infrastructure and service businesses started taking these customer concerns as one of their top priorities in their present and future product and market development activities. Hence, information security is already a household term that will stay with us forever.

We will use digital multimedia as the underlying application in *Data Hiding Fundamentals and Applications*. Data hiding has numerous multimedia applications, spanning from broadcast monitoring to fingerprinting. On the other hand, there are many heuristic and methodic solutions proposed in the literature. Furthermore, some of these techniques are already commercialized and deployed in products. In this book, we address the various issues involved in the design of data hiding systems, particularly for secure media delivery.

Data hiding is a means of secret communications using subliminal (secret) channels. Therefore, the methodology and framework emphasized in this book borrows significantly from information and communications theories in the open literature. Although we address both synchronous and asynchronous data hiding scenarios, we lay more emphasis on the synchronous data hiding case.

The target audience of this book includes graduate students in electrical engineering, computer engineering, computer science, and applied mathematics who have some familiarity with information theory and communications. The book will also benefit practicing engineers in the information security field.

Chapter 1 presents a brief introduction to data hiding and its applications. The basic concepts and terminology of the subject matter are defined. In Chapter 2, a theoretical framework for data hiding is introduced in detail. Popular data hiding methods are also reviewed in this chapter. Chapter 3 discusses the concept of communications, with side information introduced by Claude E. Shannon, and relates it to the data hiding problem. We also revisited the dualities between communications and data hiding concepts and operators in Chapter 3. Chapter 4 deals with Type I (Linear) data hiding techniques. Data hiding in transform domain is examined and performances of different linear transforms are also compared in this chapter. Nonlinear (Type II and Type III) data hiding methods are introduced in Chapter 5. The performance comparisons of Type III data hiding scenarios using different post-processing techniques are presented. Chapter 6 delves into the data hiding applications and interconnects the theoretical framework introduced in the early chapters with the practical concerns in the real world. Major design issues and some advanced solutions in data hiding systems are covered in Chapter 7. In this chapter, we also highlighted a set of data hiding attacks that a practical system needs to be robust against. Data hiding applications that are used in a secure multimedia delivery system are presented in Chapter 8. Although we tried to be as thorough as possible in the references cited in the book, there still might be omissions since the field is still evolving. A bibliography follows Chapter 8.

The authors of the book have performed research in data hiding since 1996. Ramkumar and Akansu also had the opportunity to spend several years in the industry to extend and implement their theoretical knowledge in a secure peer-to-peer (P2P) multimedia delivery platform. They also had the privilege to interact with prominent researchers in the relevant fields, as well as the successful business people who build products and services around these technological advances. The authors acknowledge that this book would not have happened without their invaluable collaborations, support, and contributions.

Husrev T. Sencar
Mahalingam Ramkumar
Ali N. Akansu

Newark, New Jersey
July 2004

Introduction

The rapidly decreasing cost of processing, storage, and bandwidth has already made digital media increasingly popular over traditional analog media. However, digital media also causes extensive vulnerabilities to mass piracy of copyrighted material. It is, therefore, very important to have the capabilities to detect copyright violations and control access to digital media. Fueled by these concerns, data hiding has evolved as an enabler of potential applications for copyright protection such as access control of digital multimedia (e.g., watermarking), embedded captioning, secret communications (e.g., steganography), tamper detection, and others.

1.1   What Is Data Hiding?

Data hiding is the art of hiding a *message signal* in a *host signal* without any perceptual distortion of the host signal. The composite signal is usually referred to as the *stego* signal. Data hiding is a form of *subliminal* communication. Any form of communication relies on a channel or medium. Data hiding, or steganographic, communications rely on the channel used to transmit the *host content*. As the stego content moves around the globe, perhaps over the Internet, or by any other means usually deployed for communicating the host signals, so does the *embedded*, covert message signal.

Traditional Communication Scheme

Communication by Data Hiding

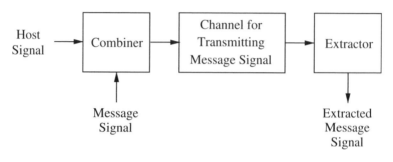

Figure 1-1 Traditional communication and steganographic communication block diagrams.

Figure 1-1 displays the block diagram for steganographic communication as compared to traditional communication systems.

1.2   Forms of Data Hiding

Data hiding may be classified in different ways based on

   (1) the relative importance of cover and message signals,
   (2) the nature of cover content,
   (3) the need for an unmodified cover signal for message signal extraction,
   (4) the type of subliminal communication—synchronous or asynchronous, and

(5) the type of attacks to remove hidden messages—active and passive wardens.

1.2.1   Relative Importance of Cover Signals

If the cover signal is of little or no importance, the stego signal may even be generated synthetically. In other words, the only purpose of the stego signal is subversive communication of the message signal (the underlying cover signal is of no relevance). This technique is common in *linguistic steganography* [1]. For example, the following message, sent by a German spy in World War II [2],

Apparently neutral's protest is thoroughly discounted and ignored. Isman hard hit. Blockade issue affects pretext for embargo on byproducts, ejecting suets and vegetable oils,

translates to

Pershing sails from NY June 1

if we take the second letter in each word.

Consider another example in which the cover signal serves a purpose. The Vedas, ancient Hindu literature, contain numerous instances of data hiding. Sanskrit, the language of the Vedas, assigns a number between 0 and 9 for all consonants. By a particular choice of consonants, and utilizing the freedom in the choice of vowels, one can compose poetic hymns with many interpretations. Here is an example of a sutra of spiritual content as well as mathematical significance [3]:

gopi bhagya madhuvrata
srngiso dadhi sandhiga
khala jivita khatava
gala hala rasandara.

This stego signal translates to the cover signal

O Lord anointed with the yogurt of the milkmaids' worship,
O savior of the fallen,
O master of Shiva, please protect me.

In addition, this verse yields a hidden message signal—the value of $\frac{\pi}{10} =$ 0.31415926535897932384626433832792 to 32 decimal places.

For the type of data hiding described in this book, the cover content is relatively more important. It is in fact for the protection of such cover content that we purport to use data hiding.

1.2.2 Nature of Content

Another type of classification for data hiding depends on the nature of the cover signal. We have signals like text or binary data that are *inherently discontinuous*. On the other hand, we have signals like audio and images that are *inherently continuous* (even though we may handle digitized versions of them). This book addresses data hiding only for the latter type of signals (inherently continuous).

More precisely, signals A_1 and A_2 are inherently continuous if $A_1, A_2 \in \mathcal{A}$, $\mathcal{A}$ metric space with distance measure d; and if $d(A_1, A_2)$ is "small," A_1 and A_2 are "perceptually close." Obviously, text or binary data do not satisfy the above condition, because even if one bit is changed in an executable binary code, the difference in interpretation may be significant.

1.2.3 Oblivious and Nonoblivious

Yet another classification of data hiding may depend on whether the original unmodified content is needed for the extraction of the hidden message signal. Nonoblivious data hiding methods need the original, while oblivious methods do not. We shall see that both methods are used for secure multimedia applications. This book addresses both types of data hiding methods, even though oblivious methods are treated in more detail.

1.2.4 Synchronous and Asynchronous

Data hiding is a form of communication, and communication schemes can be of synchronous or asynchronous type. In conventional communication systems, an unknown propagation delay is the main reason that both carrier and symbols have to be synchronized. For communications scenarios employing data hiding, the carrier is the content. Obviously, the message-carrying content could undergo far more types of distortions than the carrier signals for conventional communication systems. Simple and commonly used operations like resampling the content can cause a loss of synchronization during detection. Therefore, the problem of achieving synchronization for data hiding is more challenging than traditional carrier and symbol synchronization methods.

It is also possible for data hiding to be completely asynchronous. However, we shall see that similar to the case in conventional communication schemes, such methods cannot be as efficient as synchronous communications. In this book, we focus more on synchronous data hiding. However, we also address some methods to regain synchronization once it is lost.

1.2.5 Active and Passive Wardens

This classification is based on "interceptors" of the stego signals in the channel that may modify the stego signal. For the case of steganography with *active wardens*, we have active interceptors in the channel trying to sabotage the secret communication (perhaps by introducing modifications to the stego signal with the idea of thwarting the attempted secret communication).

On the other hand, if the stego signal passes unmodified through the channel (or, if any modification occurs, it is not for the purpose of destroying the hidden message signal), we have steganographic communication with a *passive warden*. We shall consider both types of steganographic techniques in this book.

1.3 Properties of Steganographic Communications

Applications of data hiding emerge from the fact that steganographic communications have some rather unique properties:

(1) The original or cover signal is modified. Data hiding would not be possible if we were not able or not permitted to modify the original signal.

(2) The modification introduced to the signal persists throughout the remaining life of the signal.

(3) The modification introduced by data hiding is imperceptible, or the original (cover) signal and the modified (stego) signal should be perceptually indistinguishable.

(4) Preexisting channels are utilized. Communication of the covert message does not demand a channel of its own. The cover signal is used as the transmission medium.

(5) Secret communication techniques like cryptography conceal the transmitted message. Steganography conceals the very fact that some form of communication is being attempted.

1.3.1 Multimedia Data Hiding

Data hiding in multimedia [4], [5], [6], [7] can help in providing proof of the origin and distribution of a content. Multimedia content providers can communicate with the *compliant multimedia players* through the *subliminal*, steganographic channel. This communication modality might control or restrict access of multimedia content and carry out e-commerce functions for pay-per-use implementations. The concept of compliant multimedia players may extend to computer operating systems that would recognize protected multimedia files, meaning one may not be able to print a document or make additional copies unless authorized by the hidden data in the document. Note that all material available on paper may eventually be in electronic form. Downloading or distributing the documents could be controlled by the hidden data in the documents.

A typical application scenario of data hiding for multimedia content delivery is depicted in Fig. 1-2. The *content providers* supply the raw multimedia data (e.g., a full-length movie) along with some hidden agents, or *control data*. The job of the *distributors* would be to package the content in some suitable format (like MPEG [Motion Picture Experts Group]) consumable by the players and distribute the multimedia either through DVDs/CDs, live digital broadcasts, or even by hosting websites for downloads. The compliant multimedia *players* will typically be connected to the Internet.

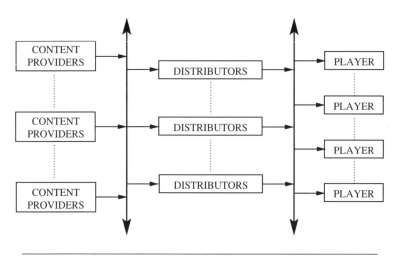

Figure 1-2   A secure multimedia delivery system.

In conventional multimedia distribution, the content provider loses all control over how the multimedia is used at the moment it leaves his/her hands. The key idea behind employing data hiding in this scenario is to reestablish control whenever the content is used. The content provider, by hiding some agents in raw data, hopes to control access to the multimedia content. This can be done with the *cooperation of the players* and an *established protocol for communication* between the *content providers* and the *compliant multimedia players*.

Perhaps a more conservative secure media distribution scheme, Fig. 1-3, would also rely on encryption of distributed content in addition to using data hiding for tracing origin and destinations of content. Multimedia content would be subjected to watermarking for the purpose of embedding ownership information and possibly some additional control information (like a pointer to a URL). This would then be followed by data compression, encryption, and perhaps some form of channel coding before the content is delivered to various consumers. The consumers equipped with compliant players could obtain decryption keys, extract and honor the control information, and use data hiding to insert the fingerprint of the consumer before the decoded content is sent in the clear to the renderer. If the rendered content is captured and redistributed, such content

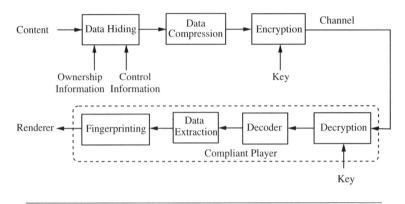

Figure 1-3   Secure media distribution scheme.

would carry the fingerprint of the compliant player where the content was decrypted.

1.4   The Steganographic Channel

Figure 1-4 depicts a block diagram of a general data hiding channel. $\mathbf{C}$ is the original multimedia data, which is also referred to as the *cover* signal. The cover signal serves as the carrier for the hidden message m. The message

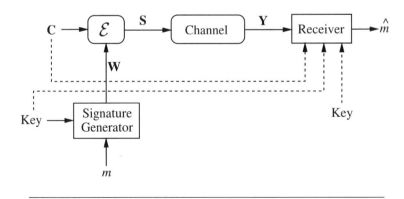

Figure 1-4   Block diagram of multimedia steganography.

m is converted to a signature $\mathbf{W}$ that is in a form suitable for embedding in $\mathbf{C}$ as expressed in

$$\mathbf{W} = \mathcal{W}(m, \mathcal{K}) \tag{1.1}$$

where $\mathcal{W}$ is the signature generator block in Fig. 1-4 and $\mathcal{K}$ is a key. Usually the embedding operation $\mathcal{E}$ takes the form of superpositioning of $\mathbf{W}$ with $\mathbf{C}$ to obtain $\mathbf{S}$. However, other forms of embedding are also possible. Therefore,

$$\mathbf{S} = \mathbf{C} + \mathbf{W} \quad \text{or} \quad \mathbf{S} = \mathcal{E}(\mathbf{C}, \mathbf{W}). \tag{1.2}$$

The *imperceptibly modified* multimedia signal $\mathbf{S}$ (the *stego* signal) is transmitted through a channel $\mathcal{C}$ and emerges at the other end as $\mathbf{Y} = \mathcal{C}(\mathbf{S})$. Now, the buried message signal m is retrieved by a detector $\mathcal{D}$ as

$$\hat{m} = \mathcal{D}(\mathbf{Y}, \mathcal{K}). \tag{1.3}$$

Note that the detector is assumed to have the key $\mathcal{K}$ available. For nonoblivious data hiding, (e.g., for watermarking applications), the detector $\mathcal{D}$ may also require the original $\mathbf{C}$ for extracting the hidden message or signature. Hence,

$$\hat{m} = \mathcal{D}(\mathbf{Y}, \mathbf{C}, \mathcal{K}). \tag{1.4}$$

The channel, apart from other things, may include a lossy compressor at one end and decompresser at the other end. While this is the main cause of concern for most data hiding applications, the same is not true for watermarking applications. For watermarking applications, the channel may include agents with *deliberate intentions* of removing the watermark.

Frameworks for Data Hiding

A tremendous amount of information signals are transmitted every second through a wide variety of media—radio, television, telephones, and, of course, the ubiquitous Internet. Such information signals can be transmitted by different means. For example, a local television channel may be received by a cable station through the air and then rebroadcast to homes through cables. In many applications, it is of great interest for the originators of such information to attach an auxiliary message signal to their content (or information signals) in order to monitor its moves. Data hiding, as we saw in the previous chapter, is a mechanism for attaching an auxiliary (message) signal to such information (cover) signals. The auxiliary message signal is *inextricably* tied to the information signal and persists independently of the means by which the information signal is disseminated. In the context of information hiding, the information signal is called the *cover signal* before the message signal is embedded into it and the *stego signal* after the message signal is embedded.

In a conventional communications scenario, an information signal typically modulates a carrier signal. The nature of the carrier signal and the mechanism of modulation may, however, vary significantly from application to application. From the point of view of a communications engineer, data hiding can be seen as *modulation of the information signal* by the *auxiliary message signal*. The fundamental reason for this is to ensure that extraction of the message signal from the information signal does not depend on *how the information signal was transmitted,* as long as the information signal is recovered with reasonable fidelity at the receiver. The fundamental difference between conventional communication scenarios and data hiding is that while modulation of the carrier signal by the information signal typically changes the carrier signal drastically, modulation of the information signal by the message signal should introduce only imperceptible changes into the information signal.

Let $c(t)$ be some information (cover) signal, and $w(t)$ the auxiliary message signal. A modulator $\mathcal{E}$ generates the composite signal $s(t) = \mathcal{E}(c(t), w(t))$, with some $d(c(t), s(t)) \leq \epsilon$, where d is a suitable distortion metric for the information signal. At the receiver, we typically have a noisy version $y(t) = s(t) + z(t)$ of the modulated information (stego) signal $s(t)$. The term $z(t)$ represents noise in the communication channel that was used to disseminate the information signal. The receiver should be able to obtain an estimate $\tilde{w}(t)$ of the auxiliary message signal $w(t)$. Obviously, we would like to obtain as faithful an estimate as possible of $w(t)$ from $y(t)$, in the presence of channel noise $z(t)$, while simultaneously ensuring that modulation of the information signal with the message signal does not alter the information signal drastically, $d(c(t), s(t)) \leq \epsilon$.

The most important issues that arise in the study of data hiding techniques concern:

- *Embedding and Detecting Mechanism.* What is the optimum way to embed and then later extract this information? In other words, what should be the nature of $\mathcal{E}, \mathcal{D}$?
- *Capacity.* What is the optimum amount of data that can be embedded in a given signal? In the above model, the capacity is related to the fidelity of the extracted data $w(t)$ or $\tilde{w}(t)$.
- *Robustness.* How do we embed and retrieve data such that it would survive malicious or accidental attacks at removal?
- *Transparency.* How do we embed data such that it does not perceptually degrade the underlying content?

These questions have been the focus of intense study in the past few years, and some remarkable progress has already been made. However, there are still more questions than answers in this rapidly evolving research area. Perhaps a key reason for this is the fact that data hiding is inherently a multidisciplinary topic that builds on developments in diverse subjects. The areas that contribute to the development of digital watermarking include at least the following:

- Information and communication theory
- Decision and detection theory
- Signal processing and transforms
- Cryptography and cryptographic protocols

Each of these areas deals with a particular aspect of the data hiding problem. Generally speaking, information- and communication-theoretic methods deal with the data embedding (encoder) side of the problem. For example, information-theoretic methods are useful in the computation of the amount of data that can be embedded in a given signal subject to various constraints such as peak power (square of the amplitude) of the embedded data or the embedding-induced distortion. The host signal can be treated as a communication channel, and various operations such as compression/decompression, filtering, etc., can be treated as noise. Using this framework, many results from classical information theory are successfully applied to compute the data hiding capacity of a host signal.

Decision theory is used to analyze data embedding procedures from the receiver (decoder) side. For a given data embedding procedure, how do we extract the hidden data from the host signal, which may have been subjected to intentional or unintentional attacks? The data extraction procedure must be able to guarantee a certain amount of reliability. What are the chances that the extracted data is indeed the original embedded data? In data hiding applications in which the embedded data is used for copyright protection, decision theory is used to detect the presence of embedded data. In applications like fingerprinting, detection-theoretic methods are needed to extract the embedded information.

A variety of signal processing tools and algorithms may be used in the field of digital watermarking. Such algorithms are based on aspects of the human visual system, properties of signal transforms (e.g., Fourier and discrete cosine transform [DCT]), noise characteristics, properties of various signal processing attacks, etc. Depending on the nature of the application and the context, some of these methods might be implemented

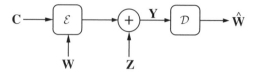

Figure 2-1   Illustration of data hiding from a signal processing perspective.

at the encoder, at the decoder, or at both. The user has the flexibility to mix and match different techniques depending on the algorithmic and computational constraints of the application.

Watermarking is a generic signal processing tool that has to be utilized efficiently to realize the desired application. Cryptography and cryptographic protocols play an important role in ensuring that watermarking is used in a proper manner to achieve content tracking in a secure information delivery system.

2.1   Signal Processing Framework

It is more common to consider the cover (host), stego, message signals, and channel noise as sampled real-valued signals expressed as $\mathbf{C}, \mathbf{S}, \mathbf{W}, \mathbf{Z} \in \Re^N$, rather than continuous real-valued signals $c(t), s(t), w(t)$ and $z(t)$.

Consider a host signal $\mathbf{C} \in \Re^N$ and a message signal $\mathbf{W} \in \Re^N$ (see Fig. 2-1). An embedder $\mathcal{E}$ embeds the message signal $\mathbf{W}$ in the host signal $\mathbf{C}$ to yield the stego signal $\mathbf{S}$ given as

$$\mathbf{S} = \mathcal{E}(\mathbf{C}, \mathbf{W}). \tag{2.1}$$

Let d be a predefined metric. In other words $e = d(\mathbf{S}, \mathbf{C})$ is the "distance" between $\mathbf{S}$ and $\mathbf{C}$. A commonly used metric or distance measure is the mean square error (MSE), given by

$$\text{MSE} = d(\mathbf{S}, \mathbf{C}) = \sum_{i=1}^{N} \frac{(\mathbf{S} - \mathbf{C})^2}{N}. \tag{2.2}$$

The *embedding distortion* $d(\mathbf{S}, \mathbf{C})$ is constrained to be less than a defined threshold P to ensure that the cover signal $\mathbf{C}$ and the stego signal $\mathbf{S}$ are perceptually the same or very similar.

The stego signal is corrupted by a noise signal $\mathbf{Z} \in \mathfrak{R}^N$ before it reaches the detector $\mathcal{D}$. The detector obtains an estimate $\widehat{\mathbf{W}} \in \mathfrak{R}^N$ of the message signal $\mathbf{W}$ defined as

$$\widehat{\mathbf{W}} = \mathcal{D}(\mathbf{Y}). \tag{2.3}$$

The problem now boils down to the optimal design of embedder $\mathcal{E}$ and detector $\mathcal{D}$ to maximize the "fidelity" of $\widehat{\mathbf{W}}$, subject to the distortion constraint $d(\mathbf{S}, \mathbf{C}) \leq P$.

An intuitive way to maximize the fidelity of $\widehat{\mathbf{W}}$ is to minimize the error signal

$$W_e = \min_{\forall a} \sum_{i=0}^{N} (a\widehat{W}[i] - W[i])^2 \tag{2.4}$$

which is the same as maximizing the correlation

$$\rho = \frac{\mathbf{W}^T \widehat{\mathbf{W}}}{|\mathbf{W}||\widehat{\mathbf{W}}|}. \tag{2.5}$$

In the case of continuous time signals, the goal would be the optimal choice of $\mathcal{E}, \mathcal{D}$ to maximize

$$\rho = \frac{\int w(t)\tilde{w}(t)\, dt}{\sqrt{\int w^2(t)\, dt \int \tilde{w}^2(t)\, dt}}. \tag{2.6}$$

2.2 Data Hiding from a Communications Perspective

We now have a host signal $\mathbf{C} \in \mathfrak{R}^N$ and a message signal $m \in \mathcal{M}$, where $\mathcal{M}$ is an alphabet of size M (see Fig. 2-2). The message signal m can equivalently be considered as an index $1 \leq m_i \leq M$ in the alphabet.

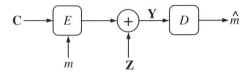

Figure 2-2   Illustration of data hiding from a communications perspective.

The encoder E produces the stego content $\mathbf{S}$ as

$$\mathbf{S} = E(\mathbf{C}, m_i) \tag{2.7}$$

subject to the distortion constraint $d(\mathbf{S}, \mathbf{C}) \leq P$.

$\mathbf{Y}$ is the corrupted version of $\mathbf{S}$ from which the decoder estimates the message signal as

$$\hat{m} = D(\mathbf{Y}). \tag{2.8}$$

For the case of continuous time signals,

$$s(t) = E(c(t), m_i) \tag{2.9}$$

and

$$\hat{m} = D(y(t)) \tag{2.10}$$

where $y(t) = s(t) + z(t)$.

The problem is now the optimal design of encoder E and decoder D to minimize

$$m_e = \sum_{i=1}^{M} \sum_{j=1, \, j \neq i}^{M} p(m_i | m_j) p(m_i) \tag{2.11}$$

where $p(m_i | m_j)$ is the conditional probability of m_i given that m_j was embedded.

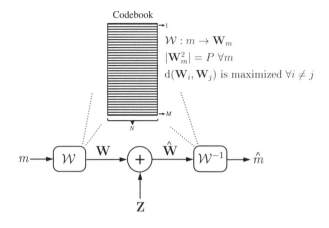

Figure 2-3 A sampled baseband spread spectrum communication scheme.

2.3 Relationship Between Communications and Signal Processing Frameworks

Consider a digital baseband spread spectrum communications scheme $(\mathcal{W}, \mathcal{W}^{-1})$ (Fig. 2-3) in which an arbitrary symbol indexed by m: $1 \leq m \leq M$ from an alphabet $\mathcal{M}$ is mapped to a sequence $\mathbf{W} \in \Re^N$. The sequence $\mathbf{W}$ is in turn transmitted over a channel characterized by an additive noise $\mathbf{Z} \sim \mathcal{N}[0, \sigma_Z^2 I]$. The corrupted vector at the receiver is $\mathbf{Y} = \mathbf{S} + \mathbf{Z}$. The mappings involved are expressed as

$$\mathcal{W} : m \rightarrow \mathbf{W} \in \Re^N \qquad \mathcal{W}^{-1} : \widehat{\mathbf{W}} \rightarrow \hat{m} \tag{2.12}$$

and may be performed by using an $M \times N$ codebook C at the transmitter and the receiver. The transmitted vector $\mathbf{W}$ is one of the possible M codewords of the codebook. The codebook should be chosen such that the codewords, $\mathbf{W}_i$, $1 \leq i \leq M$, are *maximally separated* and $\sum_{j=1}^{N} \mathbf{W}_i^2(j) = NP \; \forall i$ is satisfied. The receiver determines the element of the codebook that is "closest" to the corrupted vector $\widehat{\mathbf{W}} = \mathbf{W} + \mathbf{Z}$ in order to obtain an estimate of the transmitted symbol.

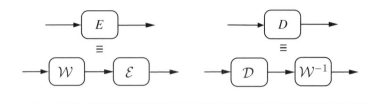

Figure 2-4   Splitting of E and D into two independent parts.

Let us now assume that the encoder E and decoder D can be separated into two independent parts as shown in Fig. 2-4. In other words, the encoder E first maps the message m to a codeword $\mathbf{W}$, which is in turn embedded in $\mathbf{C}$ by the embedder $\mathcal{E}$. Similarly, the decoder D now includes a detector $\mathcal{D}$ that obtains $\widehat{\mathbf{W}}$, an estimate of the transmitted codeword $\mathbf{W}$. Then, $\mathcal{W}^{-1}$ maps the estimated codeword $\widehat{\mathbf{W}}$ to $\hat{m}$, an estimate of the symbol m.

For the continuous-time case, for instance, the M message symbols may be mapped to M signals $w_i(t), 1 \le i \le M$, such that

$$\int w_i^2(t)\, dt = \int w_j^2(t)\, dt \quad \forall i, j \tag{2.13}$$

where

$$\rho_{ij} = \sum_{\forall i,j, i \ne j} \int w_i(t) w_j(t)\, dt \tag{2.14}$$

is as small as possible.

Therefore, the overall hiding and detection scheme illustrated in Fig. 2-5 can be described as

$$\begin{aligned} \mathbf{W} = \mathcal{W}(m) \quad \mathbf{S} = \mathcal{E}(\mathbf{C}, \mathbf{W}) \quad &\text{embedding,} \\ \widehat{\mathbf{W}} = \mathcal{D}(\mathbf{Y}) \quad \hat{m} = \mathcal{W}^{-1}(\widehat{\mathbf{W}}) \quad &\text{detection.} \end{aligned} \tag{2.15}$$

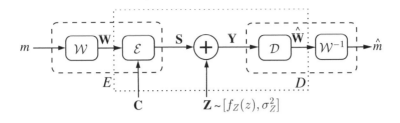

Figure 2-5   The data hiding channel with constrained embedder and detector.

Figure 2-6   Equivalent communication channel.

As far as the signaling scheme $(\mathcal{W}, \mathcal{W}^{-1})$ is concerned, the entire dashed box in Fig. 2-5 appears as a channel that can be characterized by some additive noise. This is illustrated in Fig. 2-6. Therefore, the optimal design of E and D is now split into two independent problems:

(1) the design of $(\mathcal{W}, \mathcal{W}^{-1})$
(2) the design of $(\mathcal{E}, \mathcal{D})$.

Note that the optimal design problem of $(\mathcal{E}, \mathcal{D})$ is the same as the one addressed in Section 2.1. However, as we shall see in the coming chapters, for some data hiding schemes it may not be possible to split the detector into two independent parts. In other words, for such schemes

$$\hat{m} = D(\mathbf{Y}) \tag{2.16}$$

and

$$\hat{m} = D(y(t)) \tag{2.17}$$

for the continuous case.

2.4 A Review of Data Hiding Methods

The early works in the literature on data hiding focused mainly on heuristic approaches. As the similarities between the issues of data hiding and other fields became evident, a variety of approaches were made available by exploiting those similarities. Among these approaches, the ones that have generated a lot of attention are inspired by spread spectrum communications and communication with side information [8], [9], [10], [11], [12].

Data hiding techniques are characterized by the embedding and extraction techniques employed. Methodologically, the proposed embedder/detector designs can be categorized into two main groups: additive spread spectrum–based methods and quantization-based methods.

In additive spread spectrum methods, the watermark signal is generated by modulating the information symbols with a weighted unit energy spreading vector and then is added to the host signal [13], [14], [15], [16], [17]. By choosing an appropriate weighting factor, perceptual intactness of the host signal is retained. These methods are preferable due to their ease in processing and their reliability under additive noise interference. With additive embedding, the data hiding rate is uncompromisingly traded off against its robustness to severe attacks while complying with the perceptual constraints. The major drawback of such methods is that the host signal affects as a source of interference at the detector. As a result of this fact, satisfactory performance is not possible unless the host signal is available during detection or host signal interference is negligibly smaller than channel interference. In additive schemes, optimal decoding of the watermark signal depends on exact probabilistic characterization of the host signal at the detector.

The shortcomings of additive spread spectrum methods in suppressing the host signal interference are handled by adopting the results of communication with side information to data hiding applications. Costa, in [18], showed for the first time that for an additive white Gaussian noise (AWGN) channel with Gaussian input and side information, the channel capacity does not depend on the side information. His results, when evaluated within a data hiding context, encouraged researchers in designing practical oblivious data hiding schemes that could achieve the hiding capacity. This approach gives the information *hider* a freedom in determining the rate vs robustness characteristics at a given attack level. However, the distortion measure on the host signal is confined to the squared error distance (power limited), irrespective of the perceptual features of the host signal.

In practice, in order to achieve the hiding rates that are closer to the upper capacity bound, several implementations that utilize this approach are proposed in the literature [9], [19], [20], [21], [22], [23]. These techniques are characterized by the use of enhanced quantization procedures in order to design embedding/detection methods that approximate the performance of Costa's optimal encoding/decoding. Within this approach, the optimal implementation requires higher-dimensional quantization for embedding. In [24], Zamir *et al.* show that nested lattices can be used to construct optimum codes. However, a satisfactory performance is also achievable through scalar quantization or unidimensional lattices. The extraction of the hidden message is usually achieved by employing minimum distance decoding due to the use of lattice structures in embedding. As a consequence of such an embedding, these methods are vulnerable to signal scaling. Therefore, they perform well only if the attack is not severe. However, they are suitable for oblivious data hiding applications.

Chen *et al.* [25] provide a formal treatment of data hiding methods that use quantization indices to embed signals called quantization index modulation (QIM). In this class of methods, quantization is used to force the host signal coefficients to take desired values depending on the information signal to be embedded. Similarly, Chou *et al.* in [11], [23], based on a duality with the distributed source coding problem, implement the exhaustive codeword generation for Costa's scheme by using a robust optimization method through the use of trellis coded quantizers. In this research direction, the most popular embedding technique is a low complexity implementation of QIM that relies on uniform scalar quantization, which is called dither modulation (DM) [26]. In fact, the earliest data hiding methods [27], [28], [29], [30], which modified only one or two least significant bits (LSBs) of the host signal, were based on the same principle in rejecting the host signal interference, so-called low bit modulation (LBM). For example, a method which modifies only two LSBs may be considered a form of QIM in which the step size of the quantizer used is 4. Even–odd modulation is another embedding technique that operates similarly. In the data hiding scheme proposed by Wang *et al.* [31], the significant wavelet coefficients are modified such that they *quantize* to an even or odd value depending on the bit to be embedded. In [32], Wu *et al.* introduced a similar scheme based on JPEG (Joint Photographic Experts Group) quantization by altering the DCT coefficients.

The additive spread spectrum- and quantization-based methods have poor performances for the "no attack" and "severe attack" cases, respectively. In the former, the performance becomes independent of the additive attack level, whereas in the latter the performance drops rapidly with the increase in the attack. These deficiencies point to a nonoptimal design procedure compared with Costa's scheme, which can deliver perfect host signal interference rejection at all attack levels. The need for a class of practical methods by which the hider has better control over the operating characteristics has been immediately recognized by various researchers.

In quantization-based methods, this effort has resulted in the incorporation of embedding quantization with a postprocessing function and redundancy coding of the watermark signal. In [9] and [33], Chen *et al.* introduced, respectively, a distortion-compensated (DC) version of QIM (DC-QIM) (which can achieve the capacity under AWGN attacks) and a spread transform (ST) technique for practical implementations (which embeds the watermark signal by spreading it over many host signal coefficients). Ramkumar *et al.* [21], considering scalar embedding, used a continuous triangular periodic function for extracting the watermark signal and also employed a thresholding type of processing at the embedder. Eggers *et al.* [22] optimized the performance of DC-DM by a more careful optimization of embedding parameters. They also combined multilevel signaling with binary coding techniques for low attack applications and provided some performance results [34], [35]. Perez-Gonzalez *et al.* [36] proposed a probability density function (pdf) transformation type of processing for embedding. Furthermore, they provided a calculation of an upper bound for the probability of error in a multidimensional embedding case considering various noise distributions.

In order to improve the performance of additive spread spectrum methods, a similar approach to quantization-based methods is also developed. Inspired by ST-DM, [36] proposed a decoding technique that integrates the underlying principles of quantization-based methods with the additive schemes. In this method, a watermark signal is selected such that when the linear correlation between the watermark signal and the undistorted stego signal is quantized, the resulting signal is a centroid of the lattice associated with the embedded signal. The probability of error performance of this method is improved by further processing. Consequently, the watermark signal is selected such that rather than the quantized correlation metric itself, the properly scaled error due to quantization of the correlation metric is mapped to the desired centroid. Similarly, in [37] the watermark signal energy is properly shaped to compensate for the host signal interference at the detector. This is achieved by designing the weighting as a function of the projection of the host signal onto the spreading sequence so that at the detector, the host signal's effect is diminished.

Communication with Side Information and Data Hiding

Shannon [38] introduced the first analysis of discrete memoryless channels with side information, in the form of varying channel states from a finite set, causally known to the encoder. He proved that this channel is equivalent (in terms of capacity) to a usual memoryless channel that has the same output alphabet and an expanded input alphabet with no side information. Accordingly, each letter of the new input alphabet is generated as a mapping from the set of states into the input alphabet of the original channel. Kusnetsov *et al.* [39] examined a practical version of the same problem in which the errors in the channel were invariant (namely memory with defective cells). They offered an encoding scheme for reliable storage of information when the encoder is given the defect information, and they investigated the redundancy bounds for such codes. Gelfand *et al.* [40] considered a similar channel as in [38] by removing the causality condition on the encoder such that at any transmission time, the encoder had the whole channel state information for all times. They proceeded to derive the capacity of this channel assuming an input alphabet $\mathcal{X}$, an output alphabet $\mathcal{Y}$, an auxiliary alphabet $\mathcal{U}$, and a finite set $\mathcal{C}$ of side information, where $\mathcal{X}, \mathcal{Y}, \mathcal{U}, \mathcal{C} \in \Re^N$. The channel capacity C_0

is expressed in terms of random variables $X \in \mathcal{X}$, $Y \in \mathcal{Y}$, $U \in \mathcal{U}$, and $C \in \mathcal{C}$ by a maximization over all conditional joint probability distributions $p_C(c)p_{U,X}(u,x|c)p_Y(y|x,c)$ as

$$C_0 = \max_{p(u,x|c)} (I(U,Y) - I(U,C)) \qquad (3.1)$$

where $p_X(x)$ is the probability mass function of a random variable X, and $I(X,Y)$ is the mutual information between two random variables X and Y. Heegard *et al.* [41], also using this formulation, extended the idea to establish achievable storage rates for memory when defect information is given only to the encoder or the decoder and completely to the decoder but partially to the encoder.

Costa [18] applied the results of [40] to memoryless channels with discrete time and continuous alphabets and presented an information-theoretic analysis of a problem that also applies to oblivious data hiding. He studied a communications scenario in which the encoder transmitted a message index to the decoder in the presence of side information and designed the auxiliary variable in Gelfand's formulation as $U = X + \alpha C$, where X is the power constrained input (codeword), C is the channel state information available at the encoder, and α is a scaling factor. Costa showed that for an AWGN channel with Gaussian input and side information, the channel capacity does not depend on the side information.

Later research gained considerable momentum, first by reinterpreting these results in terms of oblivious data hiding, and later by formulating the problem from a game-theoretic perspective. The researchers in [42] and [43] assumed a Gaussian distributed host signal and squared error distortion measure and studied the problem as a data hiding game between the hider-extractor and the attacker. In [42], Moulin *et al.* introduced an information-theoretic model for data hiding considering memoryless attacks. In their model, the information hider determines the embedding strategy without knowing the attack, whereas the attacker uses the stego signal to design the attack. The extractor, on the other hand, is assumed to be in a position to learn the strategy of the attacker. It is shown that for the squared error distortion measure and white Gaussian distributed host signal, the Gaussian test channel is the optimal attack, and the hiding capacity is the same as in the case of the host signal's being known to the detector. They also showed that Costa's results were valid for this setting of the data hiding game under the small distortions scenario, which assumes that host signal power is much higher than that of the distortions introduced by the hider and attacker. Cohen *et al.* [43] presented a detailed discussion and the results of hiding capacity assuming a Gaussian distributed host signal and squared error distortion measure, similar to [42], except for the removal of the assumption that the extractor knows the attack. They showed that an independent, identically distributed (*iid*) Gaussian host signal maximizes the hiding capacity among all finite fourth-moment distributions for the host signal. It is also discussed that additive attacks are suboptimal. Furthermore, they extended Costa's results by considering non–white-noise attacks and non-Gaussian embedding distortions.

These studies have shown that the solution for the hiding capacity varies with the setting of the game, and Costa's framework yields the upper bound on the coding capacity among all versions of the game, since the attacker has a fixed strategy (additive noise) that is known to both the encoder and the decoder. Therefore, Costa's framework and his results serve as a test-bed for comparing and evaluating the performances of various practical embedding/detection techniques.

3.1   Costa's Framework

Costa in [18], based on the results of [40], considers a power-constrained AWGN channel with *iid* Gaussian input $\mathbf{X}$ and side information $\mathbf{C}$ (in the form of channel state) that is available *only* at the encoder in a noncausal manner. A message index m is transmitted to the receiver by properly selecting the codeword $\mathbf{X}$ that is distorted during transmission by the additive channel state $\mathbf{C}$ and the channel noise $\mathbf{Z}$. Consequently, the channel output is defined as $\mathbf{Y} = \mathbf{X} + \mathbf{C} + \mathbf{Z}$. Considering the design of $\mathbf{U} = \mathbf{X} + \alpha\mathbf{C}$, $0 < \alpha < 1$, and assuming that $\mathbf{X}$, $\mathbf{C}$, $\mathbf{Z}$ are *iid* length N sequences of random variables with zero covariance matrices and Gaussian marginal distributions (i.e., $X \sim \mathcal{N}(0, P)$, $C \sim \mathcal{N}(0, \sigma_C^2)$, $Z \sim \mathcal{N}(0, \sigma_Z^2)$), the communication rate is computed as [18]

$$
\begin{aligned}
R(\alpha) &= I(U, Y) - I(U, C) \\
&= H(X + C + Z) - H(X + C + Z|X + \alpha C) \\
&\quad - H(X + \alpha C) + H(X + \alpha C|C) \\
&= H(X + C + Z) + H(X) - H(X + C + Z, X + \alpha C) \qquad (3.2)
\end{aligned}
$$

where $H(X)$ is defined as the entropy of random variable X. Since X, C and Z are assumed to be independent Gaussian random variables, $X + \alpha S$ and $X + C + Z$ are respectively distributed as $\mathcal{N}(0, P + \alpha^2 \sigma_C^2)$ and $\mathcal{N}(0, P + \sigma_C^2 + \sigma_Z^2)$. The joint distribution of $X + C + Z$ and $X + \alpha C$ is also Gaussian, with the density function given as

$$
\begin{aligned}
&f_{X+C+Z, X+\alpha C}(x + c + z, x + \alpha c) \\
&= \mathcal{N}\left(\begin{bmatrix} 0 \\ 0 \end{bmatrix}, \begin{bmatrix} P + \sigma_C^2 + \sigma_Z^2 & P + \alpha \sigma_C^2 \\ P + \alpha \sigma_C^2 & P + \alpha^2 \sigma_C^2 \end{bmatrix} \right).
\end{aligned} \qquad (3.3)
$$

Hence, the rate in Eq. (3.2) is obtained by calculating the entropies for the corresponding distributions as [44]

$$
R(\alpha) = \frac{1}{2} \log_2 \frac{P(P + \sigma_C^2 + \sigma_Z^2)}{P \sigma_C^2 (1 - \alpha)^2 + \sigma_Z^2 (P + \alpha^2 \sigma_C^2)}. \qquad (3.4)
$$

Maximizing $R(\alpha)$ over α, Costa shows that the communication rate achieves $\frac{1}{2} \log_2(1 + P/\sigma_Z^2)$ bits per transmission for $\alpha^* = P/(P + \sigma_Z^2)$, which is the capacity of the same AWGN channel with the side information available to both encoder and decoder. Thus, for a properly chosen α, the lack of side information at the decoder does not reduce the capacity.

The channel model for Costa's framework is displayed in Fig. 3-1. In order to transmit message m, encoder E generates the codeword $\mathbf{X}$ that is additive to the channel state $\mathbf{C}$ at the given channel noise variance. Decoder D, not knowing the random channel state $\mathbf{C}$, detects the message $\hat{m}$ from the received signal $\mathbf{Y}$.

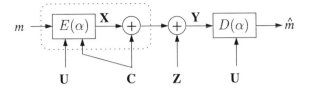

Figure 3-1 The channel model for Costa's framework corresponding to codebook design of $\mathbf{U} = \mathbf{X} + \alpha\mathbf{C}$.

Costa outlines the capacity-achieving encoding/decoding scheme based on random coding techniques. The optimal codebook has $M = \lfloor 2^{NR} \rfloor$[1] codewords corresponding to M messages. Each message is transmitted in N uses of the channel. For optimal encoding and decoding, $2^{N(I(U,Y)-\epsilon)}$ (for an arbitrarily small ϵ) length-N *iid* sequences with individual distributions $\mathcal{N}(0, P + \alpha^{*2}\sigma_C^2)$ are generated and then partitioned into 2^{NR} bins. Each bin is associated with the index of a message and points to $2^{N(I(U,C)+\epsilon)}$ sequences. This collection of sequences is made known to both encoder and decoder. In order to generate the codeword, the side information $\mathbf{C}$ is weighted by the proper α and subtracted from the sequences in the bin corresponding to the message to be conveyed. Among the resulting signals, the one that is orthogonal to $\mathbf{C}$ ($|(\mathbf{U}_j - \alpha^*\mathbf{C})^T\mathbf{C}| < \delta$, $j = 1, \ldots, 2^{N(I(U,C)+\epsilon)}$, for a proper δ value) and also satisfies the power constraint ($\frac{1}{N}\|\mathbf{X}\|^2 \leq P$) is the optimal codeword corresponding to the message index being sent.

The encoder sends the codeword over the channel. The decoder receives the signal $\mathbf{Y}$ and searches over all $\mathbf{U}$ sequences for the jointly typical $(\mathbf{U}_j, \mathbf{Y})$ pair ($|(\mathbf{U}_j - \alpha\mathbf{Y})^T\mathbf{Y}| < \delta, j = 1, \ldots, 2^{N(I(U,Y)-\epsilon)}$). The sent message is decoded successfully from the $\mathbf{U}_j$ sequence and the received signal $\mathbf{Y}$, for $\alpha = \alpha^*$ and large N, as

$$|(\mathbf{U}_j - \alpha\mathbf{Y})^T\mathbf{Y}| = |(\mathbf{U}_j - \alpha^*\mathbf{C} - \alpha^*\mathbf{X} - \alpha^*\mathbf{Z})^T(\mathbf{X} + \mathbf{C} + \mathbf{Z})| \quad (3.5)$$

$$= |(1 - \alpha^*)\mathbf{X}^T\mathbf{X} - \alpha^*\mathbf{Z}^T\mathbf{Z}| \quad (3.6)$$

$$= (1 - \alpha^*)NE[X^2] - \alpha^*NE[Z^2] \quad (3.7)$$

---

[1] $\lfloor x \rfloor$ is the greatest integer smaller than or equal to x.

$$= N \left(1 - \frac{P}{P + \sigma_Z^2}\right) P - \frac{NP}{P + \sigma_Z^2} \sigma_Z^2 = 0. \qquad (3.8)$$

The message index associated with the bin that contains the sequence U_j is declared as the sent message. Such a code generation is asymptotically optimal as $N \to \infty$ [18].

3.2 A Framework Based on Channel Adaptive Encoding and Channel Independent Decoding

For the same communications scenario, let the channel model of Costa's framework be modified in two respects. The first modification is by redefining the channel input as $\mathbf{X}_n = \mathbf{X} - \mathbf{X}_t$. We refer to $\mathbf{X}_t$ as the "processing distortion," since it is by nature a "disturbance" to encoder output $\mathbf{X}$. The processing distortion $\mathbf{X}_t$ may be a function of the encoder output $\mathbf{X}$, and the nonzero correlation between $\mathbf{X}$ and $\mathbf{X}_t$ is denoted by ρ. Also, $\mathbf{X}_t$, like $\mathbf{X}$, is *iid* and independent of $\mathbf{C}$. In this channel model, since the codeword transmitted by the encoder is $\mathbf{X}_n$, the power constraint that needs to be satisfied by the codeword $\mathbf{X}$ in Costa's framework applies to $\mathbf{X}_n$, viz., $\frac{1}{N}\|\mathbf{X}_n\|^2 \leq P$. Consequently, the received signal at the decoder is expressed as $\mathbf{Y} = \mathbf{X}_n + \mathbf{C} + \mathbf{Z}$. The second modification is by designing the shared variable as $\mathbf{U} = \mathbf{X} + \mathbf{C}$, where the α value employed in codebook generation is set to 1 regardless of the channel's noise level. This setting will be referred to as CAE-CID (channel adaptive encoding and channel independent decoding) framework.

The transmission rate for the modified channel can now be computed for $U = X + C$, $X_n = X - X_t$, and $Y = X_n + C + Z$ as

$$
\begin{aligned}
R &= I(U,Y) - I(U,C) \\
&= I(X+C, X_n+C+Z) - I(X+C,C) \\
&= H(X_n+C+Z) - H(X_n+C+Z|X+C) - H(X+C) + H(X+C|C) \\
&= H(X_n+C+Z) - H(Z-X_t|X+C) - H(X+C) + H(X) \\
&= H(X) + H(X_n+C+Z) - H(Z-X_t, X+C). \qquad (3.9)
\end{aligned}
$$

The formulation given in Eq. (3.9) can be solved for rate R assuming that random variables X, X_t, C, and Z are mutually independent except for the known dependence between X and X_t and that they are distributed according to $\mathcal{N}(0, \sigma_X^2)$, $\mathcal{N}(0, \sigma_{X_t}^2)$, $\mathcal{N}(0, \sigma_C^2)$, and $\mathcal{N}(0, \sigma_Z^2)$, respectively. The normalized correlation between X and X_t is defined as

$$\rho = \frac{E[XX_t]}{\sqrt{E[X^2]E[X_t^2]}}. \tag{3.10}$$

On the other hand, X_n is a random variable, with the second moment set to P, and its distribution depends on how X_t is related to X. Furthermore, the random variables $Z - X_t$ and $X + C$ are jointly Gaussian, with the probability density function given by

$$f_{Z-X_t,X+C}(z - x_t, x + c) = \mathcal{N}\left(\begin{bmatrix} 0 \\ 0 \end{bmatrix}, \begin{bmatrix} \sigma_Z^2 + \sigma_{X_t}^2 & E[XX_t] \\ E[XX_t] & \sigma_X^2 + \sigma_C^2 \end{bmatrix}\right). \tag{3.11}$$

Consequently, the rate in Eq. (3.9) is derived by computing the entropies for the marginal and joint distributions as [44]

$$R(\sigma_X, \sigma_{X_t}, \rho) = \frac{1}{2}\log_2\left(\frac{\sigma_X^2(P + \sigma_C^2 + \sigma_Z^2)}{(\sigma_X^2 + \sigma_C^2)(\sigma_{X_t}^2 + \sigma_Z^2) - E[XX_t]^2}\right). \tag{3.12}$$

Using Eq. (3.10), Eq. (3.12) can be rewritten as

$$R(\sigma_X, \sigma_{X_t}, \rho) = \frac{1}{2}\log_2\left(\frac{\sigma_X^2(P + \sigma_C^2 + \sigma_Z^2)}{(\sigma_X^2 + \sigma_C^2)(\sigma_{X_t}^2 + \sigma_Z^2) - \rho^2\sigma_X^2\sigma_{X_t}^2}\right). \tag{3.13}$$

The achievable transmission rate for this channel can be found by maximizing the rate R over σ_X, σ_{X_t}, and ρ under the constraint $\frac{1}{N}\|\mathbf{X} - \mathbf{X}_t\|^2 = P$. Since ρ is a normalized variable, it does not depend on the variances of X and X_t. Hence, setting $\rho = 1$ (X_t is a linear function of X) will maximize Eq. (3.13) in ρ. Moreover, the power constraint on the input relates σ_X and σ_{X_t} as

$$
\sigma_{X_t} = \begin{cases} \sigma_X - \sqrt{P}, & \text{if } \rho = 1 \\[2ex] \rho\sigma_X - \sqrt{\sigma_X^2(\rho^2 - 1) + P}, & \text{if } \rho \neq 1. \end{cases} \tag{3.14}
$$

As a result, maximization of the rate given in Eq. (3.13) reduces to a maximization over σ_X for $\rho = 1$ and $\sigma_{X_t} = \sigma_X - \sqrt{P}$. Then,

$$
\max_{\sigma_X} R\left(\sigma_X, \sigma_{X_t} = \sigma_X - \sqrt{P}, \rho = 1\right) = \frac{1}{2}\log_2\left(1 + \frac{P}{\sigma_Z^2}\right)\Bigg|_{\sigma_X = \sigma_X^*} \tag{3.15}
$$

which is maximized for

$$
\sigma_X^* = \frac{P + \sigma_Z^2}{\sqrt{P}}, \qquad \sigma_{X_t}^* = \frac{\sigma_Z^2}{\sqrt{P}}. \tag{3.16}
$$

This is the capacity of the AWGN channel in which the side information is also known to the decoder, as first derived by Costa [18]. The preceding results show that the optimal codebook design in Costa's framework based on a particular α^* can be equivalently achieved in the CAE-CID framework with the corresponding σ_X^* when $\rho = 1$. Therefore, the two frameworks are equivalent, and they can be translated into each other through $\sigma_X^* = \sqrt{P}/\alpha^*$ at the same transmission rate. The corresponding channel model for the proposed CAE-CID framework is displayed in Fig. 3-2. When compared with Fig. 3-1, the main difference is that α dependency of the (E, D) pair is replaced by the inclusion of $\mathbf{X}_t$ generated by the processing $\mathcal{P}$ at the encoder.

The optimal encoding/decoding scheme of the CAE-CID framework is similar to the one described in [18]. The codebook used for encoding and decoding relies on the design of $\mathbf{U} = \mathbf{X} + \mathbf{C}$, as α is set to unity. Correspondingly, the shared $\mathbf{U}$ sequences are *iid* with an underlying marginal

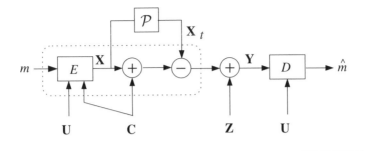

Figure 3-2 The channel model for the proposed CAE-CID framework corresponding to the codebook design of $\mathbf{U} = \mathbf{X} + \mathbf{C}$.

distribution $\mathcal{N}(0, P + \sigma_C^2)$. The channel dependence, however, is reflected in the appropriate choice of processing that generates $\mathbf{X}_t$ from $\mathbf{X}$. At the encoder, for the given $\mathbf{C}$, the jointly typical $(\mathbf{U}, \mathbf{C})$ pair is searched in the bin corresponding to the message signal being sent. The codeword is generated from the $\mathbf{U}_j$ sequence that satisfies the orthogonality constraint $(|(\mathbf{U}_j - \mathbf{C})^T \mathbf{C}| < \delta, j = 1, \ldots, 2^{N(I(U,C)+\epsilon)})$ and yields codeword $\mathbf{X}_n$ such that the power constraint $(\frac{1}{N}\|\mathbf{X}_n\|^2 \leq P)$ is satisfied. It should be noted that in order to achieve capacity, $\mathbf{X}_t$ is a linear function of $\mathbf{X}$. Therefore, the codeword $\mathbf{X}_n$ is readily obtained from the encoder output $\mathbf{X}$ by the relation $\mathbf{X}_n = (\sqrt{P}/\sigma_X)\mathbf{X}$.

On the decoder side, the sent message is decoded as the index of the bin that contains the $\mathbf{U}$ sequence, which is jointly typical with the received signal $\mathbf{Y}$. The particular sequence $\mathbf{U}_j$ is found, for large N, as

$$|(\mathbf{U}_j - \mathbf{Y})^T \mathbf{Y}| = |\left(\mathbf{U}_j - (\mathbf{X} - \mathbf{X}_t + \mathbf{C} + \mathbf{Z})\right)^T (\mathbf{X} - \mathbf{X}_t + \mathbf{C} + \mathbf{Z})|,$$

$$= |\mathbf{X}_t^T \mathbf{X} - \mathbf{X}_t^T \mathbf{X}_t - \mathbf{Z}^T \mathbf{Z}| \tag{3.17}$$

$$= NE[XX_t] - NE[X_t^2] - NE[Z^2] \tag{3.18}$$

$$= N\frac{P + \sigma_Z^2}{\sqrt{P}}\frac{\sigma_Z^2}{\sqrt{P}} - N\frac{(\sigma_Z^2)^2}{P} - N\sigma_Z^2 = 0 \tag{3.19}$$

where $E[XX_t] = \sigma_X^* \sigma_{X_t}^*$, Eq. (3.10) for $\rho = 1$, is used. The cancellation of the terms in Eq. (3.19) completely relies on the choice of $\mathbf{X}$ and the corresponding $\mathbf{X}_t$ at the encoder.

In CAE-CID framework, since the design of the shared variable is fixed as $\mathbf{U} = \mathbf{X} + \mathbf{C}$, the optimal encoding and decoding relies merely on the statistics of the encoder output $\mathbf{X}$ and its dependence with processing distortion $\mathbf{X}_t$.

3.2.1 Highlights of the CAE-CID Framework

When compared with Costa's framework, the CAE-CID framework has the following characteristics:

(1) In Costa's framework, the channel adaptive operation of encoder/decoder is achieved through proper selection of the scaling factor α. However, in the CAE-CID framework, the channel-dependent nature of the encoding is reflected in both inputs $\mathbf{X}$ and $\mathbf{X}_t$. Thus, channel state interference rejection at the decoder is achieved solely by the encoder's ability to properly select σ_X and σ_{X_t} depending on the given σ_Z, Eq. (3.16).

(2) When the channel noise level changes in the CAE-CID framework, the encoder/decoder can continue successful operation at a lower or higher rate by adjusting P at the encoder without updating the shared collection of $\mathbf{U}$ sequences, as long as

$$\sigma_X^2 \geq 2\hat{\sigma}_Z \tag{3.20}$$

where $\hat{\sigma}_Z^2$ is the new channel noise power (derivation details are given in Appendix A).

(3) The CAE-CID framework provides a better theoretical basis for practical embedder/detector designs, as the postprocessing employed in practical methods can be represented by the processing distortion term $\mathbf{X}_t$ in the formulations. In Chapter 5, practical embedder/detector designs will be studied from this point of view. Different types of postprocessing and their performances will be evaluated based on the choice of X and X_t.

3.3   On the Duality of Communications and Data Hiding ——— Frameworks

The theory of data hiding has been developed mainly through employing analytical tools of *communication with side information* and *spread spectrum communications*. This is achieved by reinterpreting and adapting basic concepts such as channel, side information, and power constraints within the context of data hiding.

In data hiding, the channel is the medium between the hider and the extractor, and it includes all forms of disturbances that affect the *stego* signal, which is an intelligent combination of the host signal and the message to be conveyed. Side information available at the encoder in a communication channel model is associated with the host signal at the embedder in the equivalent data hiding model. Similarly, the encoder/decoder pair (E, D) is functionally equivalent to the embedder/detector pair $(\mathcal{E}, \mathcal{D})$. Power constraints in a channel communication scenario are analogous to the perceptual distortion limits that are determined based on the features of the host signal. The bandwidth is dual to embedding signal size, as they are both resources of the communication, and the measure of signal-to-noise ratio (SNR) corresponds to the measure of embedding-distortion to attack-distortion ratio (watermark-to-noise ratio [WNR]). Table 3-1 shows the duality between the communications and data hiding frameworks.

Based on the frameworks given in Sections 3.1 and 3.2, encoding and decoding of a message index relies on proper selection of the codeword.

TABLE 3-1
Duality Between Communications and Data Hiding Frameworks

COMMUNICATIONS FRAMEWORK	DATA HIDING FRAMEWORK
Side information	Host signal
Encoder/decoder	Embedder/detector
Channel noise	All forms of modification on the stego signal (attack)
Power constraints	Perceptual distortion limits
Bandwidth	Embedding signal size
Signal-to-noise ratio	Embedding distortion to attack distortion ratio

Correspondingly, in the dual data hiding problem, the performance of an embedding and detection technique depends on the underlying codeword generation scheme. Hence, the main goal of a data hiding method is to design practical codebook and codeword generation schemes that can deliver perfect host signal interference rejection at all noise levels.

A codebook is a collection of mappings from the set of messages to be conveyed. Each mapping, or codeword, is generated from the host signal by an intelligent process based on the imposed distortion constraints and the expected noise level. However, in the formulations of data hiding, a codeword is defined in two different ways. From the communications point of view, the side information is a state of the channel and the codeword is the signal transmitted through the channel. Then, due to the analogy with the communications framework, a codeword can be defined as the distortion introduced into the host signal due to the embedding operation. However, within the context of data hiding, side information is the host signal, and it is also transmitted through the channel. Correspondingly, one can define the stego signal to be the codeword, as it is the channel input. In order to better exploit the duality between the communications and data hiding frameworks, the former definition for codeword is adopted.

A typical data hiding system can be modeled as

$$
\begin{aligned}
\text{Embedding:} \quad & \mathcal{W}: m \to \mathbf{W}, \\
& \mathbf{S} = \mathcal{E}(\mathbf{C}, \mathbf{W}) \\
\text{Attack:} \quad & \mathbf{Y} = \mathbf{S} + \mathbf{Z} \\
\text{Detection:} \quad & \hat{m} = \mathcal{D}(\mathbf{Y}) \quad \text{or} \quad \widehat{\mathbf{W}} = \mathcal{D}(\mathbf{Y}), \quad \mathcal{W}^{-1}: \widehat{\mathbf{W}} \to \hat{m}
\end{aligned}
$$

where the detector is assumed to have no access to the host signal during the extraction process. In this model, m is the message to be hidden, $\mathbf{C}$ is the host signal, $\mathbf{W}$ is the watermark signal, $\mathbf{S}$ is the stego signal, $\mathbf{Z}$ is the intrusion of the attacker, $\mathbf{Y}$ is the distorted stego signal, $\widehat{\mathbf{W}}$ is an estimate of $\mathbf{W}$, and $\hat{m}$ is the detected message. At the embedder, the message index m is mapped to a sequence of information samples $\mathbf{W}$ by the mapping $\mathcal{W}$, which transforms the message m into a better representation for embedding. Then, the resulting watermark signal $\mathbf{W}$ is embedded in the host signal $\mathbf{C}$. At the detector, the sent message is detected from the received signal

Y or from an extracted estimate $\widehat{\mathbf{W}}$ of **W** by the inverse mapping $\mathcal{W}^{-1}$. In the model, the embedder $\mathcal{E}$ and the detector $\mathcal{D}$ may be linear or nonlinear functions that operate on scalar or vector variables and are not necessarily inverses of each other. Not evident in the model is the distortion constraints imposed on hider and attacker for keeping the host signal intact. Ideally speaking, the measure used to quantify the hider's and attacker's distortion is expected to be in compliance with the perceptual properties of the host signal.

The frameworks in Sections 3.1 and 3.2 can be extended to data hiding by using the duality between the communications and data hiding frameworks. The encoding and decoding of both frameworks assume the presence of a very large number of **U** sequences at both the encoder and the decoder, and achieving channel capacity relies on adapting the codeword to the channel state at a given channel noise level. The encoding operation is simply a brute search in the bin pointed by the message index in order to find the **U** sequence that yields the codeword in the direction of the host signal **C**. Accordingly, each codeword is orthogonal to **C** and satisfies the power constraint P. (These constraints take the form of $\mathbf{X}^T\mathbf{C} \approx 0$ and $\frac{1}{N}\|\mathbf{X}\|^2 = P$ in Costa's framework and $\mathbf{X}_n^T\mathbf{C} \approx 0$ and $\frac{1}{N}\|\mathbf{X}_n\|^2 = P$ in the CAE-CID framework.) At the decoder, on the other hand, the same **U** sequence is searched in all bins based on joint typicality with the received **Y**. Figures 3-3 and 3-4 depict the optimal encoding and decoding for message index m. In Costa's framework, $0 < \alpha < 1$ and processing distortion is zero, whereas in the CAE-CID framework $\alpha = 1$ and the processing distortion is nonzero. Hence, the main difference between the two frameworks is in *how the channel-dependent nature is reflected in encoding and decoding operations.*

Despite its optimality, such an encoding/decoding scheme cannot be applied to the design of practical embedding/detection techniques due to complexity issues. However, its structure has been an inspiration for the design of many embedder/detector pairs [9], [19], [21], [22], [23], [36]. Common to all these data hiding techniques is the use of quantization to simplify codebook generation and codeword selection. Also, they impose the power and orthogonality constraints in a less strict sense.

An efficient algebraic structured binning scheme that generalized the approach to constructing optimum codes for data hiding is provided by

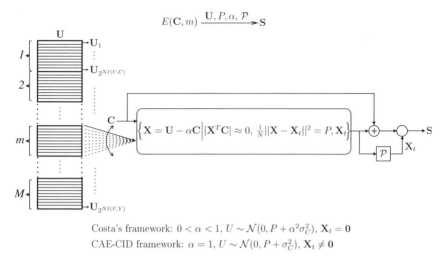

Figure 3-3 Encoding of message index m.

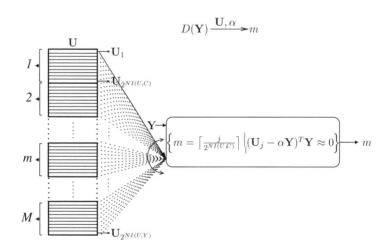

Figure 3-4 Decoding of sent message index m.

[12], [24], [45]. This approach connects the embedder/detector design problem with the areas of linear codes and lattice codes. A nested lattice code is essentially a high-dimensional lattice partition characterized by a *fine* lattice and a *coarse* lattice. The fine lattice is partitioned into a number of cosets corresponding to *coarse* lattice and its translates. (In other words, the union of the coarse lattice and its translates is the fine lattice.) Hence, the coding rate is determined by diluting the coset density in the *fine* lattice space. Accordingly, the encoding of the message m is performed by quantizing $\mathbf{C}$ in the N-dimensional signal space to the nearest lattice point in the corresponding *coarse* lattice, and the decoding is by quantizing the received $\mathbf{Y}$ to the nearest point in the *fine* lattice. Similarly, the embedding rate is designed by the number of cosets. The construction of good nested lattice codes corresponds to the use of high-dimensional vector quantization for embedding and detection. (It should be noted that QIM and DC-QIM [20] are constructions based on high-dimensional self-similar lattices where the coarse lattice is scaled and rotated version of the fine lattice.) However, from the practical point of view, high-dimensional constructions are not feasible. Therefore lattices with simpler structures need to be utilized. Such constructions include recursive quantization procedures and Cartesian products of low-dimensional lattices which coincide with the practical embedder/detector designs discussed previously.

In quantization-based methods, the optimal encoding/decoding procedure is effectively simplified by generating $\mathbf{U}$ sequences as reconstruction points, where each reconstruction point is associated with a quantizer from a set of quantizers. The number of quantizers in the set corresponds to the number of messages or message letters. Each quantizer of the set is uniquely described by a set of reconstruction points that are nonoverlapping with other sets of reconstruction points. Therefore, each finite state of $\mathbf{U}$ is a sequence with values restricted to reconstruction values of the designated quantizers. The terms $\mathbf{X}$ and $\mathbf{X}_t$ are the embedding distortion due to quantization and the processing distortion, respectively. The codeword corresponding to a message is the distortion signal introduced into the host signal as a result of the embedding operation, $\mathbf{S} - \mathbf{C}$. Consequently, it is denoted by $\mathbf{X}_n = \mathbf{X} - \mathbf{X}_t$ in the CAE-CID framework and by $\mathbf{X}$ in Costa's framework. The embedding operation, in the CAE-CID framework, is the quantization of $\mathbf{C}$ vector with the quantizer(s) pointed by the watermark signal $\mathbf{W}$ to be embedded and then the processing of the resulting quantized signal by a choice of (postprocessing) function. Hence, input $\mathbf{X}$ in the CAE-CID framework is the distortion introduced into $\mathbf{C}$ due to quantization of embedding, and the processing distortion $\mathbf{X}_t$ is the result of processing $\mathcal{P}$, $\mathbf{X}_t = \mathcal{P}(\mathbf{X})$. The sent message, on the other hand, is detected by determining the nearest reconstruction point(s) to the received signal $\mathbf{Y}$ and generating the message by mapping the corresponding quantizer(s) to the message letters they are associated with. The crux of these practical methods is that each codeword is directly generated from the given host signal and the watermark signal through quantization rather than maintaining a collection of shared $\mathbf{U}$ sequences.

Chou *et al.* in [23] applied the solution of a problem in distributed source coding to data hiding through the use of optimal quantizers. They proposed the use of robust optimization for codeword selection from Costa's huge codebook. In their work, the orthogonality of $\mathbf{C}$ and $\mathbf{X}$ is obtained by choosing $\mathbf{U}$ as a rate-distortion optimized and quantized version of a scaled version of $\mathbf{C}$. Although this approach approximates the optimal encoding and decoding scheme of Costa's framework, even the simplest implementations involve considerable complexity. Such complexity draws attention to practical approaches with simpler implementations. Chen *et al.* [9], Ramkumar *et al.* [21], Eggers *et al.* [22], and Perez-Gonzalez *et al.* [36] respectively, proposed methods to handle codebook generation by uniform scalar quantization.

3.4   Codebook Generation for Data Hiding Methods

Practical data hiding approaches can be categorized into three main types within the frameworks studied in Sections 3.1 and 3.2 based on the design of the embedder/detector pair, namely type I, type II, and type III [47], [48]. Type I methods refer to additive schemes in which the stego signal is generated by adding the watermark signal to the host signal [13], [14], [15], [16], [17]. This type of method suffers severely from host signal interference due to the nonoptimal design that assumes the host signal $\mathbf{C}$ to be a noise and tries to cancel it. Type I methods have preferable performance only if channel noise is very strong or the host signal is available at the extractor.

Type II methods are characterized by the use of quantization procedures and by the $(\mathcal{E}, D)$ pair, which are exact inverses [21], [25], [27], [28], [29], [30], [31], [32]. The major drawback of methods of this type is that they perform well only if the attack is not severe. However, they are very suitable for oblivious data hiding applications with low noise levels.

Type I and type II methods correspond to designs of $\mathbf{U} = \mathbf{X}$, $\alpha = 0$, and $\mathbf{U} = \mathbf{X} + \mathbf{C}$, $\alpha = 1$, respectively, within Costa's framework. In the CAE-CID framework, however, corresponding designs for type I and type II methods take the form $\mathbf{U} = \mathbf{X} + \mathbf{C}$ with the statistics of $\sigma_X^2 = \sigma_C^2 + \sigma_Z^2$ when $\rho = 1$, and of $\sigma_X = \sqrt{P}$ when $\mathbf{X}_t = \mathbf{0}$, respectively. These two choices of design for both frameworks correspond to two extreme cases in hiding rate vs robustness curves. Namely, type I methods are preferred for the case of severe attacks while type II methods are superior for the case of low attacks.

An optimal design is one in which the designer has control over the operating characteristics of the method. In effect, this imposes some sort of dependency on the channel noise instead of the fixed severe noise (type I) or low noise (type II) assumptions. The methods that rely on this principle are called type III which is a generalization of type I and type II. Codebook design of type III methods follows $\mathbf{U} = \mathbf{X} + \mathbf{C}$ when $\rho = 1$ ($\mathbf{X}_t \neq \mathbf{0}$) within the CAE-CID framework, and $\mathbf{U} = \mathbf{X} + \alpha \mathbf{C}$, where $0 < \alpha < 1$, within Costa's framework. Therefore, the information hider has the freedom to adapt the codeword to the host signal at the presumed noise level. These methods are ideal for oblivious data hiding.

Type III methods are developed from type II methods by enhancing the functionality of the type II embedder with added processing (i.e., thresholding, distortion compensation, and Gaussian mapping) [9], [21], [22], [36]. In type III methods, the postprocessing is designed in a way that the hiding rate is maximized for a presumed attack level [49]. However, codeword generation for most type III methods does not explicitly follow Costa's framework due to the processing that takes place after quantization of the host signal. Therefore, type III methods are better evaluated within the CAE-CID framework.

TABLE 3-2
Three Types of Embedding/Detection Schemes

	CHARACTERIZATION	CODEBOOK DESIGN
Type I	Additive schemes	$\mathbf{U} = \mathbf{X}$
Type II	Quantization-based schemes	$\mathbf{U} = \mathbf{X} + \mathbf{C}$
Type III	Channel adaptive schemes	$\mathbf{U} = \mathbf{X} + \mathbf{C}$ with processing

Table 3-2 summarizes the three types of methods. Based on the codebook designs, it is observed that type I embedding does not exploit any information on the host signal or channel noise level, while type II embedding exploits only host signal information. Type III embedding, on the other hand, utilizes both forms of information.

Figures 3-5, 3-6, and 3-7, respectively, display the codeword generation of type I, type II and type III methods for a set of watermark signals, denoted by $\mathbf{W}_1, \ldots, \mathbf{W}_M$, for the given host signal $\mathbf{C}$. In type II and type III methods, each message or watermark sample is assigned a particular quantizer $Q_\Delta(\cdot)$. The base quantizer $Q_\Delta(\cdot)$ may be a high dimensional vector quantizer or a Cartesian product of scalar quantizers with Δ as the distance between the reconstruction points. For type II embedding, $\mathbf{C}$ is quantized with respect into the watermark signal, $Q_\Delta(\mathbf{C}, \mathbf{W})$. Consequently, the codeword $\mathbf{X}$ is the quantization error introduced to the host signal $\mathbf{C}$, $\mathbf{X} = Q_\Delta(\mathbf{C}, \mathbf{W}) - \mathbf{C}$. On the other hand, in type III methods the quantization error (type II codeword) undergoes the particular processing

$$\mathcal{E}(\mathbf{C}, \mathbf{W}) \xrightarrow{P} \mathbf{S}$$

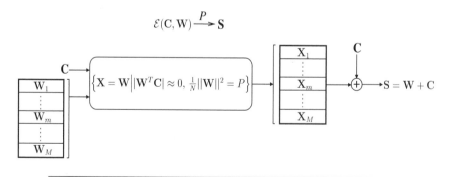

Figure 3-5   Encoding of message index m in type I methods.

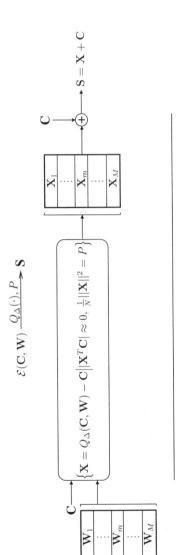

Figure 3-6   Encoding of message index m in type II methods.

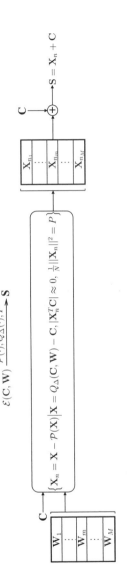

Figure 3-7   Encoding of message index m in type III methods.

$R_\times$: - - - $R_\circ$: ——

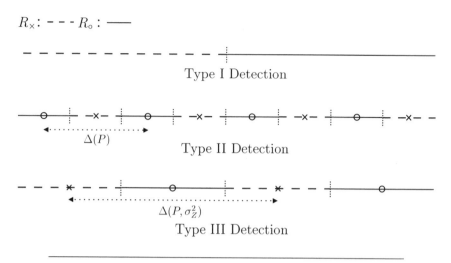

Type I Detection

$\Delta(P)$

Type II Detection

$\Delta(P, \sigma_Z^2)$

Type III Detection

Figure 3-8 The partition of the signal space between decision regions $R_\times$ and $R_\circ$ corresponding to scalar embedding and detection of a binary signal.

$\mathcal{P}$, which generates the codeword $\mathbf{X}_n = \mathbf{X} - \mathcal{P}(\mathbf{X})$. The postprocessing function $\mathcal{P}$ may have the following forms:

(1) distortion compensation [9], [22],
(2) thresholding [21], or
(3) Gaussian mapping [36].

The performance of the three types of methods can also be judged by the structure of the corresponding detectors. Considering the very simple scenario in which a two-level watermark sample is embedded in a signal coefficient and sent through a noisy channel, the three types of detectors take the following forms. The detector for the type I scheme decides on the sent sample by comparing the received signal with a threshold, whereas in type II and type III methods, detection of the embedded watermark sample is by some form of minimum distance decoding in order to determine the nearest reconstruction point to the received stego sample. Figure 3-8 displays the partitioning of the signal space between the two disjoint decision regions $R_\times$ and $R_\circ$. In the figure, $\times$ and $\circ$ denote the reconstruction points associated with the quantizers corresponding to two watermark samples. Obviously, the partitioning of the decision regions in a type I detector is far from being optimal when the channel noise level is low. This is because with a limited embedding distortion, most (host) signal coefficients are not suitable for embedding (i.e., in order to embed the information symbol denoted by o in a host signal coefficient that is at the far left of the threshold, an arbitrarily large embedding distortion needs to be introduced to translate it to the region R_o). On the contrary, the layout of the decision regions of the type II detector ensure reliable detection from all stego coefficients, albeit only up to channel distortions of power P. A type III detector, on the other hand, gives control over the size of the decision regions, and as a result successful detection can be sustained up to noise level σ_Z^2 while embedding distortion is still limited to P as in type II embedding. As the channel noise level σ_Z^2 increases, the type III detector will depart from the type II detector and take the form of a type I detector.

Figures 3-9 and 3-10 display the hiding rate vs robustness performances achievable by type I, type II, and type III methodologies computed using Eq. (3.4) for $\alpha = 0$, $\alpha = 1$, and $\alpha = P/(P + \sigma_Z^2)$, respectively, or equivalently, solving Eqs. (3.9) and (3.13) for $\sigma_X = (P + \sigma_Z^2)/(2\sqrt{P})$

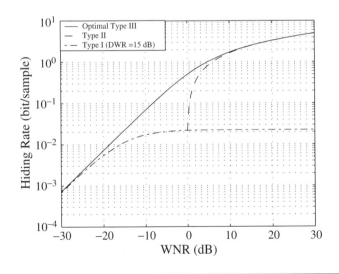

Figure 3-9 Hiding rate vs robustness performance of type I, type II, and type III methods with $P = 10$ and DWR = 15 dB.

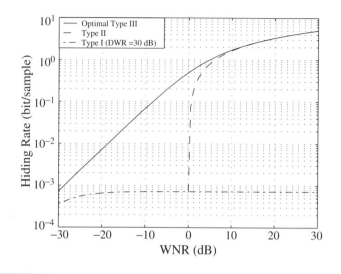

Figure 3-10   Hiding rate vs robustness performance of type I, type II, and type III methods with $P = 10$ and DWR = 30 dB.

when $\rho = 1$, $\sigma_X = \sqrt{P}$ when $\mathbf{X}_t = \mathbf{0}$, and $\sigma_X = (P + \sigma_Z^2)/\sqrt{P}$ when $\rho = 1$.

The hiding rate is measured in the number of bits that can be hidden in a host signal coefficient, and the robustness measure is defined in terms of the ratio between the embedding distortion power and the channel noise power,

$$WNR = 10 \log_{10} \frac{P}{\sigma_Z^2} \text{ in dB.} \tag{3.21}$$

However, for type I methods, the WNR by itself cannot be the indicator of robustness as the host signal is considered to be part of the noise. Therefore, another measure that can be considered is the ratio of the host signal power to the embedding distortion power,

$$DWR = 10 \log_{10} \frac{\sigma_C^2}{P} \text{ in dB.} \tag{3.22}$$

In type II methods, due to the ability to reject the host signal interference (depending on the WNR), the dependency of the performance to the document to watermark ratio (DWR) level is weak. Type I methods achieve the capacity at very low WNRs; and at high WNRs, there is almost a constant gap with the capacity. On the other hand, type II methods achieve the capacity at higher WNRs, and the hiding rate drops exponentially with the decreasing WNR. Furthermore, at low WNR, range hiding is not possible. Since type III is a superset of type I and type II methods, its optimal version can achieve the capacity at all WNRs.

A detailed analysis of type I embedding/detection and capacity results can be found in [48], [50], [51], [52].

Type I (Linear) Data Hiding

Most promising data hiding applications—like authentication, copyright control, and ownership verification—involve multimedia data and assume the presence of a very powerful attacker. Many of the data hiding techniques that are proposed for such applications are based on type I data hiding. There are two reasons for the popularity of type I methods:

(1) simplicity of implementation, and
(2) robustness to severe additive attacks.

In this chapter we investigate the achievable hiding capacities (payload) of type I (linear) data hiding in still images. For type I (linear) data hiding, the embedder $\mathcal{E}$ and the detector $\mathcal{D}$ take the forms

$$\mathbf{S} = \mathcal{E}(\mathbf{C}, \mathbf{W}) = \mathbf{C} + \mathbf{W},$$
$$\widehat{\mathbf{W}} = \mathcal{D}(\mathbf{S}) = \mathbf{S}. \tag{4.1}$$

In Eq. (4.1), the host signal $\mathbf{C}$ is an image, and the stego signal $\mathbf{S}$ is the watermarked image.

4.1   Linear Data Hiding in Transform Domain

Most of the state-of-the-art techniques for data hiding in images utilize some decomposition for embedding the message bits. Among different orthonormal decomposition techniques, it was probably the inspiration of image compression applications that caused DCT, and subband (wavelet) transforms have been more popular than the others. Another reason for the choice of DCT and wavelet-based techniques is perhaps to "match" the data hiding [53] technique with the properties of the signal processing that the image is most likely to undergo. Currently, the most common image compression tools are the DCT-based JPEG and the subband (wavelet) transform–based source coding techniques of SPIHT (set partitioning in hierarchial trees)/EZW (embedded zerotree wavelet) [54]. Adding the signature or the message signal *intelligently* in the DCT domain (for example, taking the JPEG quantization tables into account) can ensure robustness to JPEG compression. Similarly, one could design wavelet-based methods that are robust to SPIHT/EZW compression techniques. It is no surprise that most wavelet-based data hiding methods are very robust to EZW or SPIHT compression [55], although they are not very robust to JPEG compression. Similarly, DCT-based data hiding methods are robust to JPEG and not so to SPIHT/EZW compression. Of course, one cannot expect robustness of these data hiding methods to other forms of compression/signal-processing operations. Though it is true that most images are very likely to go through DCT/wavelet-transform–based compression, the situation is different for video sequences. For most video frames the major source of information is the motion vectors. Therefore, it is difficult to *intelligently* devise DCT/wavelet-transform–based methods for data hiding in video frames.

It is of great interest to devise robust data hiding methods given that no knowledge of the compression technique to be performed is available. Now the question is, what is the underlying decomposition technique that should be used? We attempt to answer that question in this chapter. We provide an information-theoretic approach to estimate the achievable capacities for different orthonormal decompositions like DCT, wavelet (subband), discrete Fourier transform (DFT), Hadamard and Hartley transforms.

Several authors, [56], [57], [58], have proposed information-theoretic approaches to characterize or evaluate the performance of the data hiding channel. In [56], Smith *et al.* model the image as a Gaussian noise source of variance defined by the average noise (image) power. The data hiding capacity (payload) is then calculated as the capacity of the Gaussian channel. In [57], Servetto *et al.* obtain the capacity of the data hiding channel where the noise source is an intentional jamming. However, it is assumed that the original image is available at the receiver. Hernandez *et al.* [58] proposed a more thorough model, which analyzes the performance of a proposed data hiding method. In this model, L orthogonal sequences are used for the signature. The image is decomposed into channels corresponding to its projections onto each of the orthogonal signatures. The capacity of the channels is calculated for unprocessed images and for images that have gone through filtering operations.

4.2   Problem Statement

Let $\mathbf{I}$ be the original (cover) image, to which a message $\mathbf{W}$ (a representation for embedded information bits) is added, such that $\hat{\mathbf{I}} = \mathbf{I} + \mathbf{W}$. The modified image $\hat{\mathbf{I}}$ is *visually indistinguishable* from $\mathbf{I}$ and may typically be subjected to a lossy compression, like JPEG, and $\tilde{\mathbf{I}} = \mathcal{C}(\hat{\mathbf{I}})$, where $\mathcal{C}(\cdot)$ denotes the compression/decompression operations pair. The embedded bits in image $\mathbf{I}$ are extracted from $\tilde{\mathbf{I}}$. We would like to know in advance the maximum number of bits that can be hidden (payload) and recovered later at the detector with an arbitrarily low probability of error; namely, the *capacity of the data hiding channel* for a given compression scenario.

A block diagram of the data hiding channel is shown in Fig. 4-1. $\mathbf{W}$ is the message (signature) to be transmitted through the channel. The channel

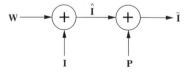

Figure 4-1   The data hiding channel.

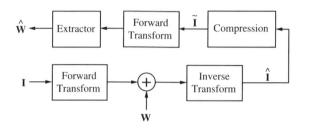

Figure 4-2   Generalized schematic of data hiding/retrieval.

has two sources of noise: $\mathbf{I}$, the noise due to the (original) cover image, and $\mathbf{P}$, the noise component due to some future processing (compression/decompression or others). $\widehat{\mathbf{W}} = \tilde{\mathbf{I}}$ is the "corrupted" or noisy message. Note that for nonoblivious data hiding methods, there is only one source of noise due to processing. The image noise can be subtracted from the received image $\tilde{\mathbf{I}}$ because it is available at the detector. One can expect such methods to have higher data hiding capacity than oblivious detection methods.

Figure 4-2 displays the block diagram of a typical data hiding method. The forward transform block decomposes the image $\mathbf{I}$ into its coefficients of L bands. A component of the signature/message signal is added to each band. The inverse transform reconstructs the modified image $\hat{\mathbf{I}}$ due to data hiding.

The image $\hat{\mathbf{I}}$ then undergoes some processing (e.g., lossy compression), resulting in the image $\tilde{\mathbf{I}}$. The hidden message signal/signature is extracted from $\tilde{\mathbf{I}}$. First, the image $\tilde{\mathbf{I}}$ is decomposed into L bands by using the same forward transform. Then, each component of the signature is extracted separately in each band. In this chapter, we assume the data hiding system shown in Fig. 4-2 and estimate its data hiding capacity for different decomposition types (different forward and inverse transform blocks).

4.3   Capacity of Additive Noise Channels

Prior to considering the data hiding channel of Fig. 4-1, we consider the simpler channel displayed in Fig. 4-3a. $X \sim [f_X(x), \sigma_x^2]$ is the message

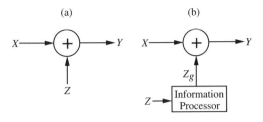

Figure 4-3   (a) A simple additive noise channel with an arbitrary noise type. (b) The channel of (a) modified to obtain its equivalent additive Gaussian noise.

signal to be transmitted, $Z \sim [f_Z(z), \sigma_z^2]$ is the additive noise in the channel, and $Y \sim [f_Y(y), \sigma_y^2]$ is the received signal at the output of the channel.

We assume that X and Z are independent. This implies that $\sigma_y^2 = \sigma_x^2 + \sigma_z^2$. Therefore, the channel capacity is expressed as [44]

$$C = \max_{f_X(x)} I_M(X, Y) = \max_{f_X(x)} h(Y) - h(Y|X) = \max_{f_X(x)} h(Y) - h(Z) \text{ bits} \quad (4.2)$$

where $I_M(X, Y)$ is the *mutual information* between X and Y. For a given noise statistic $f_Z(z)$ and input variance σ_x^2, one can maximize the entropy of the channel output Y

$$y(Y) = -\int f_Y(y) \log_2(f_Y(y)) \, \mathrm{d}y \text{ bits} \quad (4.3)$$

by choosing a suitable distribution $f_X(x)$ for the input message X. For a given variance σ_y^2, the maximum entropy value of $Y = \frac{1}{2} \log_2(2\pi e \sigma_y^2)$ bits is achieved when Y has a normal distribution. For instance, the maximum entropy value is achievable if both pdf's $f_Z(z)$ and $f_X(x)$ are normally distributed. However, for an arbitrary distribution $f_Z(z)$, and a fixed σ_x^2, the maximum achievable entropy value is not immediately obvious. To calculate this, we pass the noise Z through an ideal *information processor* (see Fig. 4-3b) that does not alter the amount of information in Z, but changes its statistics to a Gaussian distribution at its output Z_g. (The information processor can be considered an ideal data compressor, where "compression" is measured in terms of signal energy. The information processor translates the data into a form that has minimum energy while maintaining the information content or *entropy*.) Since the output of the information processor has the same entropy as its input, the variance of the output, $\sigma_{Z_g}^2$, can be obtained by solving the equation

$$h(Z_g) = h(Z) = \frac{1}{2} \log_2(2\pi e \sigma_{Z_g}^2) \text{ bits.} \qquad (4.4)$$

It is well known that Gaussian distribution has the highest entropy for a given variance [44]. Alternately, Gaussian distribution has the least variance for a given entropy. Thus, it is always true that $\sigma_{Z_g}^2 \leq \sigma_z^2$. We call $\sigma_{Z_g}^2$ the *entropy-equivalent Gaussian variance*. The maximum value of $h(Y)$ is, therefore, obtained as

$$\max_{f_X(x)} h(Y) = \max_{f_X(x)} h(X + Z_g) = \tfrac{1}{2} \log_2(2\pi e(\sigma_{Z_g}^2 + \sigma_x^2)) \text{ bits.} \qquad (4.5)$$

In order to calculate the channel capacity, we can now replace $f_Z(z)$ by $N[0, \sigma_{Z_g}^2]$ as follows

$$C = \max_{f_X(x)} h(Y) - h(Z_g) = \frac{1}{2} \log_2\left(1 + \frac{\sigma_x^2}{\sigma_{Z_g}^2}\right) \text{ bits.} \qquad (4.6)$$

Note that if the processing noise is Gaussian and independent of the image noise, the two channel noise sources in Fig. 4-1 can be replaced by a single Gaussian noise source of variance $\sigma_{I_g}^2 + \sigma_P^2$, where $\sigma_{I_g}^2$ is the equivalent Gaussian variance for the image noise **I**, and σ_P^2 is the variance of the processing noise. If σ_W^2 is the message signal energy, the capacity of the data hiding channel can be expressed as

$$C_h = \frac{1}{2} \log_2\left(1 + \frac{\sigma_W^2}{\sigma_{I_g}^2 + \sigma_P^2}\right) \text{ bits.} \qquad (4.7)$$

As a first approach to calculate the capacity or payload of the data hiding channel, the original image pixels, the image noise **I** are assumed to be uniformly distributed random variables I taking values between 0 and 255 with variance σ_I^2. Let σ_p^2 be the variance of the noise (per pixel) introduced due to processing (e.g., compression). As we shall see, the processing noise is an *estimate* of the variance of an *equivalent additive noise* that substitutes for the actual nonlinear processing noise sources (mainly quantization in the case of lossy compression). Since we do not know anything about the distribution of the equivalent processing noise, we assume the worst: Gaussian distribution. Finally, let σ_W^2 be the average energy per pixel allowed for the message signal. If MN is the number of pixels in an image, then the energy (or variance if it has a zero mean) of the message signal is calculated as

$$\sigma_W^2 = \frac{\sum_{I=1}^{MN} W_I^2}{MN} \tag{4.8}$$

where W_I is the message signal added to the Ith pixel. The (differential) entropies $h(g)$ of a Gaussian random variable g, with variance of σ_g^2, and $h(u)$, that of a uniformly distributed random variable u with variance σ_u^2, are expressed as [44]

$$h(g) = \tfrac{1}{2}\log_2(2\pi e \sigma_g^2)\, \text{bits}, \quad h(u) = \tfrac{1}{2}\log_2(12\sigma_u^2)\, \text{bits}. \tag{4.9}$$

From Eq. (4.9), the *entropy-equivalent Gaussian noise* (or the Gaussian random variable that has the same entropy as the uniform random variable u of variance σ_i^2) has a variance given by

$$\sigma_{I_g}^2 = \frac{12}{2\pi e}\sigma_I^2. \tag{4.10}$$

Although we would expect the variance of u, the pixel values, to be given by $\sigma_I^2 = \frac{255^2}{12}$ (or $\sigma_I = 73.6$), statistics from many test images (see Section 4.4 for the details of the test images used) show that $\sigma_I = 55$. Therefore, we assume that u has a uniform distribution with $\sigma_I = 55$. From Eq. (4.10) it is calculated that $\sigma_{I_g} = 55(\frac{12}{2\pi e})^{0.5} \approx 46$. If we allow a degradation of the image after the addition of a message with a peak SNR (PSNR) of up to 40 dB, then the message energy is calculated to be $\sigma_W^2 = 6.5$. Furthermore, if the image goes through a JPEG compression at 50% quality level, then it is measured for test images that the processing noise has a standard deviation of $\sigma_P \approx 6.7$ (the actual procedure for estimating processing noise is described in Section 4.4.2). This would yield a capacity C_h value of 0.0022 bits/pixel (140 bits of capacity for a 256×256 image). Even if the message-embedded image (stego image) undergoes some other processing that results in a barely recognizable image, corresponding to $\sigma_P \approx 20$, the capacity C_h would still be 0.0019 bits per pixel (about 124 bits of capacity for a 256×256 image). Therefore, one can see that hiding the message in the image domain can be very robust. However, in most cases, we do not require such robustness. Since many data hiding applications aim to protect and ascertain copyright or control access, it is unlikely in such a scenario that anyone would want to claim ownership or control access of an image of no commercial value (an image that has been significantly degraded in perceptual quality). Typically, it is sufficient if the message survives well-known image compression/decompression operations with acceptable quality.

Given that less robustness than the previously mentioned method offers is acceptable, could we do better than this? In our first approach, what we have done is very similar to the method reported in [56] (the only difference is that we have also introduced processing noise into the channel). By assuming a Gaussian channel, we imply that the image pixels have a flat spectrum. However, it is well known that the spatial frequency characteristics of a typical image is far from flat (white). Most of the image energy is concentrated in the low-frequency bands. It is, therefore, intuitive that a decomposition of an image into its different frequency bands might help. We expect the low-frequency bands of the decomposition to be very noisy due to the high-energy content of the image. On the other hand, high-frequency components would be very vulnerable to processing, as most compressors would discard them at low bit rates. At midfrequency bands, however, we could strike a compromise. A typical distribution of image and processing noise in various bands of a decomposition is shown in Fig. 4-4.

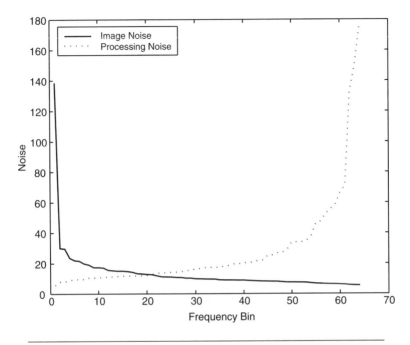

Figure 4-4 A typical distribution of image and processing noise among different bands.

In Fig. 4-5, the channel of Fig. 4-1 is decomposed into its multiple subchannels. The decomposition is performed by the forward and inverse transform blocks of Fig. 4-2. The decomposition of an image into its L subbands results in L parallel subchannels with two noise sources in each. Let $\sigma_{I_j}^2$, $j = 1, \ldots, L$, be the variances of the coefficients for each subband (or the variances of the image noise in each subchannel) of the decomposition. Similarly, let their corresponding equivalent Gaussian variances be $\sigma_{I_{g_j}}^2$. If $\sigma_{P_j}^2$ is the variance of the processing noise (Gaussian) in the jth subchannel, then the total capacity of the L parallel subchannels is given by

$$C_h = \frac{MN}{2L} \sum_{j=1}^{L} \log_2 \left(1 + \frac{v_j^2}{\sigma_{I_{g_j}}^2 + \sigma_{P_j}^2} \right) \text{ bits} \qquad (4.11)$$

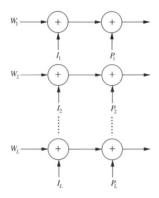

Figure 4-5   Decomposition of the data hiding channel into its parallel sub-channels (subbands).

for an image of size MN pixels. In Eq. (4.11), v_j is the *visual threshold* of band j. In other words, v_j^2 is the maximum message signal energy permitted in band j based on its perceptual quality effects. Note that if the channel was purely energy constrained (or if the constraint was on the total signature energy regardless of its distribution among different bands), then the best solution would be to use the water-filling approach [44] to calculate the overall channel capacity. However, in this case, the allowed maximum signal energy in a channel is constrained by the visual threshold of the band. Ideally, we would like to utilize all channels to the fullest extent possible.

In the following sections, we evaluate the capacity of the data hiding channel for DCT, DFT, Hadamard, and uniform-subband-decomposition–based transform domain embedding methods. We use popular compression methods like JPEG and SPIHT to model the processing (compression) noise in each subband of the decomposition.

4.4   Modeling Channel Noise

In order to model the channel noise (the two noise sources I and P in Fig. 4-1), we measure their statistics from 15 monochrome test images of size 256×256 and their JPEG and SPIHT compressed versions at various quality factors (bit rates).

4.4.1   Modeling Image Noise

The cover images are decomposed into L subbands using an orthonormal transform. Let $f_{I_j}(i_j)$ be the distribution of the jth subband with variance $\sigma_{I_j}^2$. (The image noise $\mathbf{I}$ is split into its components in L subchannels, which are modeled as random variables with $f_{I_j}(i_j)$ and variances $\sigma_{I_j}^2$, $j = 1, \ldots, L$.)

After calculating the variances of the image noise in each subchannel, the next step is to obtain the subchannels' entropy-equivalent Gaussian variances. This is achieved by plotting a histogram of the transform coefficients for each band and calculating the entropy. If Δx is the width of the n bins of the histogram $g_j(m)$, $m = 1, \ldots, n$, and p is the total number of coefficients in band j, the entropy H_j and the equivalent Gaussian variance $\sigma_{I_{g_j}}^2$ of the subband are obtained as

$$H_j = -\sum_{I=1}^{n} \frac{g_j(I)}{p\Delta x} \log_2 \left(\frac{g(I)}{p\Delta x} \right) \Delta x, \text{ bits} \quad \sigma_{I_{g_j}}^2 = \frac{2^{2H_j}}{2\pi e}.$$

Thus, the image noise in subchannel (band) j can be substituted by a Gaussian noise of variance $\sigma_{I_{g_j}}^2$. In our simulations, the image noise is estimated for each image individually for five different transforms tested.

4.4.2   Modeling Processing Noise

At the outset, one should note that processing noise is introduced due to quantization of transform coefficients. While one could accurately estimate the type of quantization noise introduced by JPEG compression on the DCT coefficients of the original image (assuming that the quantization table is known), the same cannot be feasible, for instance, for the Hadamard transform coefficients of the original image. The quantization of one DCT coefficient in JPEG compression would affect a number of Hadamard coefficients, since their subspectra are not the same. More importantly, for the reasons explained earlier, we wish to make the model of the processing noise more general. The only reason we restrict ourselves to JPEG and SPIHT image compression techniques for processing noise sources is their widespread availability. We define processing noise as the *equivalent additive noise that accounts for the reduction in correlation between the transform coefficients of the original image and the image obtained after lossy compression.* Note that while this estimate provides us with the *variance* of the equivalent additive noise, it does not tell us anything about the statistical nature of the noise (like its distribution). We therefore assume the worst—Gaussian distribution for the processing noise.

Let the processing noise variance in each subchannel be $\sigma_{P_j}^2$, $j = 1, \ldots, L$. The steps to obtain the processing noise variance in our experiments are as follows:

- Decompose the n_i original test images using an orthonormal transform.
- Obtain $\frac{MNn_i}{L}$ samples for each subband. Let i_{j_k}, $k = 1, \ldots, \frac{MNn_i}{L}$ be the coefficients of subband j.
- Apply lossy compression/decompression (JPEG/SPIHT at various quality factors) to the n_i test images.
- Decompose the n_i reconstructed images (compressed) using the same orthonormal transform.
- Let $\tilde{i}_{j_k}$, $k = 1, \ldots, \frac{MNn_i}{L}$ be the corresponding transform coefficients of the compressed images.
- Define the intraband correlation as

$$\rho_j = \frac{\langle i_j, \tilde{i}_j \rangle}{|i_j||\tilde{i}_j|} = \frac{\langle i_j, (i_j + n_j) \rangle}{|i_j||i_j + n_j|} \tag{4.12}$$

where n_j is a vector of random variables, uncorrelated with i_j.

- $\sigma_{n_j}^2 = |n_j|^2$ is the variance of the *equivalent additive noise due to compression* (or $\sigma_{p_j} = \sigma_{n_j}$).
- Since $\langle i_j, n_j \rangle = 0$, Eq. (4.12) can be simplified to obtain

$$\sigma_{P_j}^2 = |n_j|^2 = \left(\frac{1}{\rho_j^2} - 1\right)|i_j|^2. \tag{4.13}$$

It can easily be seen that the processing noise in each subband *cannot* be simply obtained as $\tilde{i}_{j_k} - i_{j_k}$. Consider a scenario, in which DCT is used for the decomposition, and a low-quality JPEG for processing. Let us assume that a high-frequency subband is completely removed due to compression ($\tilde{i}_{j_k} = 0 \; \forall k$ for some j). This implies that all information buried in that subchannel (subband) is lost. In other words, the processing noise in that subchannel has *infinite* variance (and *not* the variance of $\tilde{i}_j$). This happens because no *correlation* exists between $\tilde{i}_{j_k}$ and i_{j_k}. Note that in Eq. (4.13), when $\rho_j \to 0$, $\sigma_{P_j} \to \infty$.

Also note that while the image noise is estimated individually for each image, the processing noise is not. There are two reasons for this:

- As the equivalent image noise is estimated by correlation, the result is likely to be more accurate if more samples are used. If we calculate processing noise for each image separately (for 256×256 images using some 64-band decomposition), we have only 1024 coefficients in each band. However, using 15 images yields 1024×15 coefficients per band.

- The second reason is that this method of estimating the processing noise would yield unrealistic (very low) estimates of processing noise for low-entropy images. The original and compressed versions of low-entropy images are bound to be very "close", leading to high correlation in most bands. This would cause an overestimate of capacity for smooth images. To mitigate this effect, we average processing noise over many images.

4.5   Visual Threshold

The value of the *visual threshold* for subchannel j, v_j in Eq. (4.11), however, is highly subjective. Since the amount of message signal energy permitted in any subband is determined by the visual threshold, different models for visual thresholds would yield different estimates of achievable data hiding capacity. The visual threshold depends not only on the band, but also on the magnitude of the particular coefficient. Within the same band, a coefficient with high magnitude can be altered to a larger extent than a coefficient with small magnitude. Additionally, the visual threshold may also depend on the magnitudes of coefficients of other bands corresponding to the same image block (spatial location).

However, what we desire is an estimate of the *average* energy for the message signal that can be added to a particular band. Since it is well known that the human visual system is more sensitive to the lower frequencies than the higher ones, the SNR (message signal to image noise) should be smaller for lower-frequency subbands. In general, lower-frequency image subbands have higher variances. Hence, a reasonable model for the visual threshold factor v_j could be

$$v_j^2 = K\sigma_{I_j}^{2\alpha} \tag{4.14}$$

where $0 < \alpha < 1$, and $K \ll \sigma_{I_j}$ $\forall j$ is a constant. When $\alpha = 0$, the message signal energy is distributed equally among all subbands regardless of their variances. On the other hand, when $\alpha = 1$, the message signal energy is distributed among subchannels according to their band variances.

From Eqs. (4.11) and (4.14), for the case of *no processing noise*, if we assume that all subchannels have the same pdf type (such that $K\sigma_{I_j} = K_1\sigma_{I_{g_j}}$), the channel capacity can be calculated as

$$C_h = \frac{MN}{2L} \sum_{j=1}^{L} \log_2 \left(1 + \frac{K_1\sigma_{I_{g_j}}^{2\alpha}}{\sigma_{I_{g_j}}^2} \right) \approx \frac{MN}{2L} \log_2 \left(1 + \sum_{j=1}^{L} \frac{K_1}{\sigma_{I_{g_j}}^{2(1-\alpha)}} \right). \tag{4.15}$$

In the above equation, the approximation is justified because $(K_1\sigma_{I_{g_j}}^{2\alpha})/(\sigma_{I_{g_j}}^2) \ll 1$ $\forall j$. Note that for the case of $\alpha = 1$, the decomposition does not have any effect on the capacity. However, for $\alpha < 1$, C_h can be increased by choosing a suitable transform, as shown in the next section. Thus, the increase in capacity is due to the fact that one can add *relatively* more message signal energy to bands of lower variances (or high-frequency bands in a typical scenario).

However, in Eq. (4.14), there seems to be no rationale for fixing the value of α apart from actual simulations. We therefore adopt a different model for visual threshold. To derive the model, we argue that JPEG, at a reasonably good quality factor, is well tuned visually in distributing the quantization errors amongst the bands, at least with respect to preserving the visual fidelity of the compressed image. More advanced methods like SPIHT tend to optimize the MSE rather than visual fidelity (in general, the visual quality of a JPEG compressed image at a certain PSNR is much better than that of a compressed SPIHT image at the same PSNR). Let i_{j_k} be the coefficients of the original images and $\tilde{i}_{j_k}$ the coefficients of the same images that have gone through JPEG-75 (quality factor 75) compression and decompression. Let $\sigma_{q_j}^2$ be the variance of the quantization error, $e_{q_j} = \tilde{I}_j - I_j$, for subband j. If quantization error (due to JPEG-75) of variance $\sigma_{q_j}^2$ in subband j results in an image that is *visually satisfactory*, we can argue that the addition of a message signal with energy $\sigma_{q_j}^2$ in subband j would still render the image $\hat{\mathbf{I}}$ with an acceptable visual quality. However, in order to maintain the PSNR of $\hat{\mathbf{I}}$ in the range of 40–50 dB (so that the $\hat{\mathbf{I}}$ is visually indistinguishable from $\mathbf{I}$), we choose the subband visual thresholds as

$$v_j^2 = K_2 \sigma_{q_j}^2 \qquad (4.16)$$

where $K_2 < 1$. (The average PSNR of JPEG-75 images is only about 35 dB. Hence, a choice of $K_2 = 1$ would yield images $\hat{\mathbf{I}}$ of PSNR 35 dB. This might not be an acceptable quality. For our simulations we use $K_2 = 0.25$.)

4.6   Channel Capacity vs. Choice of Transform

It should be noted that both Eqs. (4.11) and (4.15) are subject to the following constraints:

$$\sum_{j=1}^{L} \sigma_{I_j}^2 = L\sigma_I^2, \quad \sum_{j=1}^{L} \sigma_{I_{g_j}}^2 = L\sigma_{I_g}^2, \quad \mathcal{I} = \frac{1}{2}\log_2(2\pi e \sigma_{I_g}^2)$$

where σ_I^2 is the variance of images, $\sigma_{I_g}^2$ is the entropy-equivalent Gaussian variance for σ_I^2, and $\mathcal{I}$ is the average entropy of image pixels. The first equation states that unitary transforms (the transforms used for the embedding decompositions) preserve energy. The second and third equations state that the transforms also preserve entropy. With these constraints, it can be shown that the *minimum* channel capacity (for the case of *no processing noise* or Eq. (4.15)) is achieved for $\sigma_{I_{g_j}} = \sigma \; \forall j$, or when no decomposition (spatial embedding) is used.

Note that a transform with good energy compaction or higher gain in transform coding gain (GTC) [59] would result in more *imbalance* of the coefficient variances. This would enhance the term $\sum_{j=1}^{L} K_1/(\sigma_{I_{g_j}}^{2(1-\alpha)})$ in Eq. (4.15) and therefore increase the capacity (when the processing noise is small). Hence, good energy-compaction transforms like DCT and subband transforms are good embedding decompositions for *low processing noise scenarios*.

However, the relationship between processing noise and the choice of transform is not immediately obvious. For example, if we use JPEG at low quality factor for compression and DCT as the embedding decomposition, it is very easy to see that the processing noise will approach infinity for many high-frequency bands, as they are bound to be completely eliminated. On the other hand, the high-frequency coefficients of, say, a Hadamard transform will have components in many more DCT coefficients. So, it is not very likely that any Hadamard transform band is completely eliminated. In fact, even if the processing that the image undergoes is SPIHT, it is still likely to affect the high-frequency DCT coefficients more than the high-frequency Hadamard transform coefficients, since the latter have poorer spectral selectivity. Any efficient compression method would affect the low-variance (high-frequency) bands of the transforms suitable for compression (or high-GTC transforms).

To illustrate this point, Fig. 4-6 shows the distribution of the processing noise for DCT and Hadamard transform bands for processing noise due to SPIHT compression at 1 bpp and 0.35 bpp. While the processing noise for the two decompositions are comparable for SPIHT at 1 bpp, it is seen that processing noise increases drastically for high-frequency DCT bands for SPIHT at 0.35 bpp. The high-frequency bands of the Hadamard transform, however, are relatively immune to processing noise. Similarly, low-quality JPEG affects the high-frequency bands of subband decomposition (using an 8-tap binomial QMF [Daubechies] filter) to a much larger

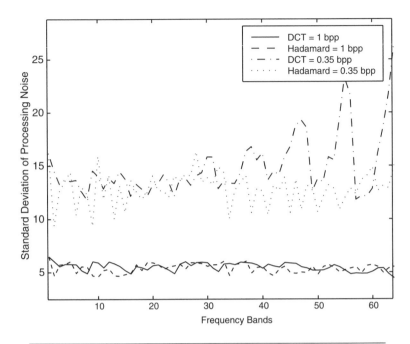

Figure 4-6   Comparison of standard deviations of processing noise for DCT and Hadamard decompositions. The source of processing noise is SPIHT compression at 1 bpp and 0.35 bpp.

extent than the high-frequency Hadamard bands. We already know that low-frequency bands are not efficient channels due to the presence of high image noise. If the high-frequency bands are also affected by processing, it leaves us a small number of useful midfrequency bands. Transforms with lower GTC have many more of these useful midfrequency bands than the high-GTC transforms due to their spectral properties at higher processing noise scenarios. Therefore, *decompositions unsuitable for compression would, in general, be more immune to processing noise than decompositions with high GTC*. Also, recall that in Section 4.3, embedding in the image domain (or using identity transform for the transform blocks in Fig. 4-2) was found to be very robust to processing noise. The identity transform, which has the lowest GTC, has the highest robustness to processing noise. It is relevant to point out here that the term *robustness* is a measure of the change in *overall capacity* with a change in the process-ing noise (or processing scenario). The more robust the decomposition, the less is the reduction in capacity for a scenario of increased processing noise (or lower-quality compression). One should note that the robust-ness of the *low-frequency bands* of, say, the DCT decomposition will be much higher than the robustness of the single band coefficients (pix-els) in the image domain. However, the low-frequency bands of the DCT have very little capacity due to high image noise. The reduced robust-ness of DCT is because of the drastic reduction in the *overall capacity* due to the significant increase of processing noise in the high-frequency bands.

The next question that arises concerns the choice of the number of bands for the decomposition or size of the transform. From Eq. (4.15), we see that a decomposition will not hurt. At worst, it may cause no improvement. Therefore, decomposing each subchannel of, for instance, a 16-band decomposition further into four subchannels can only improve the capacity of data hiding, at least when processing noise is low.

4.7   Some Capacity Results and Discussions

The estimated capacities for different 64-band decompositions (for 256×256 images, or 65,536 pixels each) like DFT, DCT, subband, Hart-ley, and Hadamard transformations are shown in Fig. 4-7. The capacities were estimated for five different transforms for eight different processing scenarios and averaged over 15 images. Figures 4-8 and 4-9 show the individual capacities of 4 different images (Baboon, Barbara, Lena, and Bridge).

Figure 4-10 shows the average channel capacities of each video frame of three video sequences (Table Tennis, Football, and Garden) averaged over 90 frames per sequence. The source of processing for the video sequences is MPEG-2 compression (30 frames/sec, 15 frames in GOP (group of pictures) and I/P (intra/predictive) frame distance of 3), at var-ious bit rates. In Fig. 4-10, the left column is the capacity estimates of I-frames and the right column is for P/B-frames.

For the subband decomposition, we use the 8-tap binomial QMF (Daubechies) filter (though it would be a better idea to use the linear phase 9-7 filters, which are used more commonly for subband or wavelet image compression, the *biorthogonality* of the filters would complicate the analysis). More specifically, we use uniform subband decomposition. For the DFT decomposition, we use only the *magnitude* of the DFT coefficients. The phase is ignored. (In other words, the message signal added would change only the magnitude of the DFT coefficients. The phase is left intact. As no message signal information is available in the phase, the phase is ignored during detection of the message signal.) The 2-D DFT of an 8×8 real matrix has 4 real and 60 complex (out of which only 30 are unique)

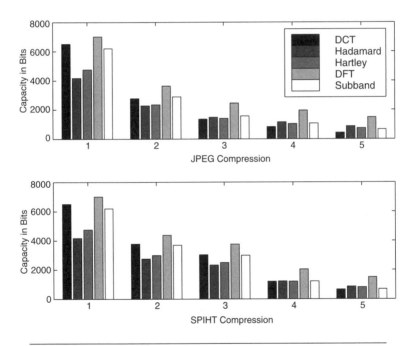

Figure 4-7 Average capacity estimates for 15 256 × 256 images. The indices for JPEG compression correspond to different JPEG quality factors (1 = lossless compression, 2 = 75%, 3 = 50%, 4 = 35%, 5 = 25%). The indices for SPIHT compression correspond to different bit rates (1 = lossless, 2 = 1 bpp, 3 = 0.75 bpp, 4 = 0.5 bpp, 5 = 0.35 bpp).

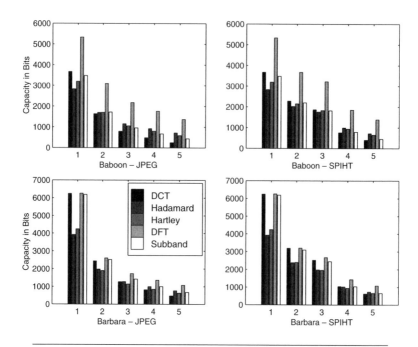

Figure 4-8 Capacity estimates for 256 × 256 Baboon and Barbara images. The indices for JPEG compression correspond to different JPEG quality factors (1 = lossless compression, 2 = 75%, 3 = 50%, 4 = 35%, 5 = 25%). The indices for SPIHT compression correspond to different bit rates (1 = lossless, 2 = 1 bpp, 3 = 0.75 bpp, 4 = 0.5 bpp, 5 = 0.35 bpp).

coefficients. Note that this causes a reduction in the number of available channels from 64 to 34, as only 34 magnitude coefficients are unique (the magnitudes of 30 complex and 4 real coefficients). In addition, this also reduces the message energy available to each channel by a factor of (approximately) half—only half the message signal energy distributed among the 60 complex coefficients is available for detection. Half the message signal energy is added just for the purpose of maintaining the symmetry properties of the DFT for a real signal. But by compromising some channels (or by reducing the degrees of freedom), we obtain smaller noise variances in each channel. As an example, consider N *iid* random variables (N degrees of freedom) with variance σ^2. If we construct $\frac{N}{2}$ random variables from the N original variables by averaging every two of them, the variance of

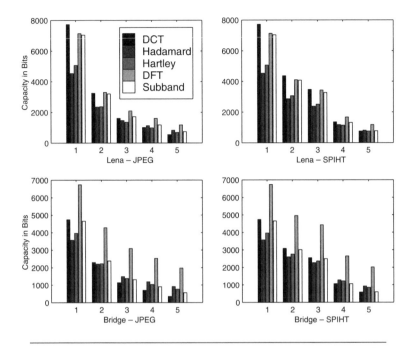

Figure 4-9 Capacity estimates for 256 × 256 Lena and Bridge images. The indices for JPEG compression correspond to different JPEG quality factors (1 = lossless compression, 2 = 75%, 3 = 50%, 4 = 35%, 5 =25%). The indices for SPIHT compression correspond to different bit rates (1 = lossless, 2 = 1 bpp, 3 = 0.75 bpp, 4 = 0.5 bpp, 5 = 0.35 bpp).

the resultant $\frac{N}{2}$ random variables will be *iid* with variances equal to $\frac{\sigma^2}{2}$. Therefore, we reduce the variance of noise in the channels by reducing the degrees of freedom (from N to $\frac{N}{2}$).

From the plots in Figs. 4-7–4-10, we see that capacities for all decompositions fall with increased processing noise as expected. DCT and subband decompositions are better than Hartley and Hadamard decompositions for detection of the message when processing noise is low. It is also seen that decompositions unfavorable for compression (DFT, Hartley, and Hadamard) are more immune to processing noise than decompositions suitable for compression (DCT, subband).

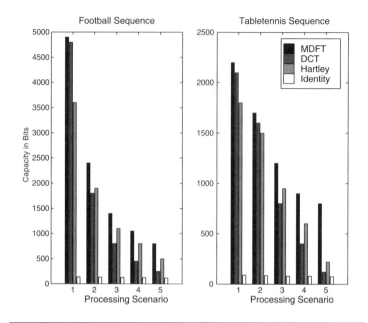

Figure 4-10   Channel capacities of different decompositions for Football and Tabletennis sequences. The processing scenarios 1–5 correspond to lossless compression and compression ratios of 10, 25, 50, and 100 (MPEG-2), respectively.

What is surprising is that magnitude DFT decomposition offers more capacity than better energy-compaction transforms even when there is no processing noise. In this case, a reduction in the entropy of the image noise is achieved by ignoring the phase of the DFT coefficients. The reduction in entropy is precisely the information content in the DFT phase. Apparently, this reduction in entropy more than offsets the reduced signal energy available for detection (again, only half the signal energy is available for detection, as the added signal power is divided among 64 coefficients, while only 34 of them are available for detection). Yet magnitude DFT performs better than other transforms because *DFT phase contains disproportionately more information than DFT magnitude!* Note that in Figs. 4-8 and 4-9, the capacity of magnitude DFT decomposition for Baboon and Bridge images is much higher than that of the high-GTC transforms, even for a no-processing-noise scenario. On the other hand, the capacity of magnitude DFT is comparable to or even less than high-GTC transforms for smoother images like Lena and Barbara. This might be due to the following reasons:

- High-GTC transforms suitable for most images are not very well suited for these high-activity images.
- The disparity between information content in the phase and magnitude is even more pronounced for these high-activity images.

In addition, being a relatively low GTC transform, DFT is also robust to processing noise like Hadamard and Hartley transforms.

Another surprising observation is that embedding in the DCT domain is slightly more resistant to subband compression methods than JPEG. Similarly, embedding in the subband domain is slightly more resistant to JPEG than SPIHT. This may appear to contradict the idea of "matching" embedding transforms with the compression method. But one should note that the matching is useful only if we design the methods *intelligently*. So, designing a DCT-based data hiding method with no idea of, say, the quantization matrix used may not be more robust to JPEG than a wavelet-based data hiding method.

As an indicator of the performance of these decompositions for other possible compression methods, we look at the capacities of the decompositions when an image has to survive JPEG *or* SPIHT. We group the four different processing scenarios of JPEG and SPIHT into four pairs: JPEG-75 and SPIHT 1 bpp, JPEG-50 and SPIHT 0.75 bpp, JPEG-35 and SPIHT 0.5 bpp, and JPEG-25 and SPIHT 0.35 bpp. For example, to calculate the capacity when the message signal has to survive JPEG-50 *or* SPIHT 0.75 bpp, we choose the worst processing noise in each subband (from the estimates of processing noise for SPIHT 0.75 bpp and JPEG-50). The capacities thus obtained are plotted in Fig. 4-11. Note that the estimates of the capacity still follow the same trend.

We can define a *figure of merit* for each of the L ($\frac{L}{2} + 2$ for magnitude DFT) subchannels for the various decompositions. The figure of merit is given as the ratio of the capacity of each subchannel to the logarithm of the power of the message signal in that subchannel. The approximate (rounded) values of the figure of merit for the channels of

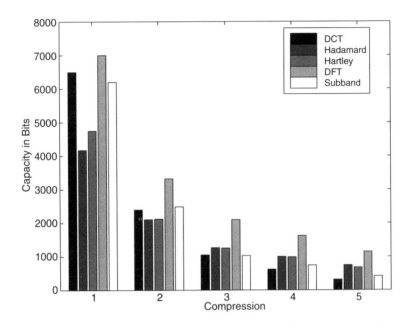

Figure 4-11   Average capacity estimates for 15 images when the message signal has to survive SPIHT *or* JPEG. The compression indices 1–5 correspond to 1 = lossless compression, 2 = (JPEG-75, SPIHT 1 bpp), 3 = (JPEG-50, SPIHT 0.75 bpp), 4 = (JPEG-35, SPIHT 0.5 bpp), 5 = (JPEG-25, SPIHT 0.35 bpp).

different decompositions (when the message has to survive SPIHT 0.5 bpp or JPEG-35) are listed in Table 4-1 for various 64-band decompositions. These figures indicate the relative performance of each subchannel and would, therefore, be useful in designing hidden communication methods to make optimal trade-offs between the visual quality of the image and the number of bits that can be embedded. As the figure of merit is normalized with respect to the message signal energy in each band, it is independent of the model used for the visual threshold. The high figures of merit for the channels of the magnitude DFT decomposition show that it would perform better than other decompositions for any message signal energy assignment method (model for visual threshold).

TABLE 4-1
Figure of Merit of the Bands of Different Decompositions When the Image Has to Survive SPIHT 0.5 bpp

(a) Magnitude DFT								(b) DCT							
0	27	49	69	83	0	0	0	0	8	19	29	37	42	29	23
27	53	72	70	87	0	0	0	8	17	28	34	41	28	10	28
49	72	69	38	51	0	0	0	19	28	36	40	35	15	7	22
69	70	38	18	32	0	0	0	29	34	40	40	23	8	2	22
83	87	51	32	43	0	0	0	37	41	35	23	15	2	11	2
0	69	46	33	0	0	0	0	42	28	15	8	2	0	0	0
0	71	69	46	0	0	0	0	29	10	7	2	11	0	0	6
0	54	71	69	0	0	0	0	23	28	22	22	2	0	6	14
(c) Uniform Subband								(d) Hadamard							
0	9	29	37	43	41	37	33	0	23	11	22	5	22	10	22
9	18	19	26	37	43	32	18	23	34	30	12	38	24	34	22
29	19	30	37	29	23	30	16	11	30	31	24	22	29	28	26
37	26	37	28	44	43	10	8	22	12	24	13	28	21	27	13
43	37	29	44	11	19	2	7	5	38	22	28	11	32	17	30
41	43	23	43	19	39	6	9	22	24	29	21	32	22	33	24
37	32	30	10	2	6	2	12	10	34	28	27	17	33	24	30
33	18	16	8	7	9	12	11	22	22	26	13	30	24	30	17

Figure 4-12 shows the average data hiding capacities for 15 images for 256-band decompositions. As expected, we observe an increase in the estimate of the capacity. The increase is more substantial for low-processing-noise scenarios.

Finally, note that we evaluate processing noise by measuring the correlation between the image components before and after compression. By this, we implicitly assume that the message signal (signature) is affected to the same extent as the image coefficients themselves by the compressor/decompressor pair. In a practical method, this may not be true. In fact, an *ideal* compression method would completely suppress any extra information added to the image coefficients (no data hiding would be possible with an ideal compression method). But practical compression methods can probably be tricked into believing that the embedded information is

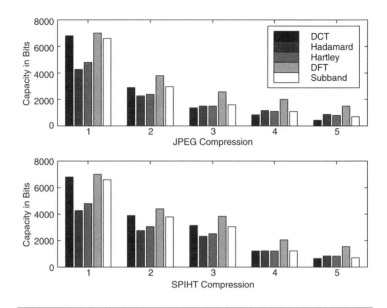

Figure 4-12   Average capacity estimates for 15 256 × 256 images for 256-band decomposition. The indices for JPEG compression correspond to different JPEG quality factors (1 = lossless compression, 2 = 75%, 3 = 50%, 4 = 35%, 5 = 25%). The indices for SPIHT compression correspond to different bit rates (1 = lossless, 2 = 1 bpp, 3 = 0.75 bpp, 4 = 0.5 bpp, 5 = 0.35 bpp).

an *integral part* of the image if the embedded message signals are chosen *intelligently*. However, choosing the signature S *intelligently* may imply reduced degrees of freedom for its choice, translating into reduced data hiding capacity.

4.8   The Ideal Decomposition

For a moment, if we ignore the magnitude DFT decomposition, the performance of a decomposition depends roughly on its position on the GTC scale. In Fig. 4-13, a few transforms are marked on the GTC scale. To the extreme left is the identity transform, which has no energy compaction. To the extreme right is the Karhunen-Loeve transform (KLT) [59]. Transforms

Figure 4-13 The GTC scale.

to the right would yield high data hiding capacities for low-processing-noise scenarios. As the processing noise increases, we should move toward the left to choose a transform. The question is, given a processing noise scenario, what would be the ideal decomposition?

For example, if $\alpha = 0.5$ in Eq. (4.14), the data hiding capacity of *each* subchannel of a decomposition is given by

$$C_{h_j} = \log_2\left(1 + \frac{K\sigma_{I_{g_j}}}{\sigma_{I_{g_j}}^2 + \sigma_{P_j}^2}\right). \tag{4.17}$$

In order to maximize C_{h_j}, it is enough to maximize $t = (\sigma_{I_{g_j}})/(\sigma_{I_{g_j}}^2 + \sigma_{P_j}^2)$. It can be easily seen that t (and hence C_{h_j}) is maximized when $\sigma_{I_{g_j}}^2 = \sigma_{P_j}^2$. The ideal decomposition would be one that results in image noise variances close to the processing noise variances in the maximum number of subbands. Typically, for high-GTC decompositions, (Fig. 4-14a), $\sigma_I \gg \sigma_P$ in the low-frequency bands and $\sigma_P \gg \sigma_I$ in the high-frequency bands. For lower-GTC transforms, the discrepancy is reduced (Fig. 4-14b). On the other hand, for the identity transform, $\sigma_I \gg \sigma_P$ in the single band (Figure 4-14c). Therefore, for the ideal decomposition, the image and processing noise variances should be distributed as shown in Fig. 4-14d. For the ideal decomposition, the image and processing noise variances should be distributed as shown in Fig. 4-14. It should also be noted that a decomposition so obtained would perform as expected only if we were able to assume the same model for the relationship between the coefficient variance and the visual threshold. Therefore, the search for such a decomposition may not be simple.

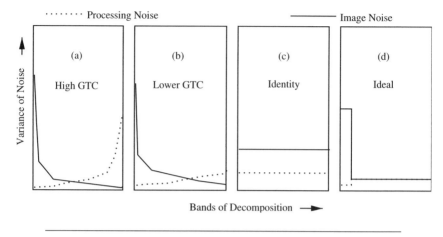

Figure 4-14 The ideal decomposition.

4.9 Factors Influencing Choice of Transform

The superiority of the magnitude DFT decomposition, among the decompositions compared, lies in an advantageous trade-off, where we compromise the degrees of freedom to reduce the entropy of the image. Simulations show that the magnitude DFT decomposition yields uniformly superior performance (over other decompositions considered) for both low- and high-processing-noise scenarios.

The final choice of decomposition should depend on the data hiding application. While some data hiding applications, like watermarking, may need robustness to intentional tampering, other applications, like captioning, may not. The performance of magnitude DFT decomposition is superior to others because of its low information content. For the very same reason, the magnitude of DFT coefficients can be altered significantly without affecting the visual quality of the image. This makes the DFT coefficients very vulnerable to intentional tampering. Thus, the magnitude DFT decomposition may not be a suitable choice for watermarking applications. However, standard image compression methods do not seem to affect the magnitude DFT coefficients drastically. This "hole" in standard compression methods can be put to use advantageously. Therefore, for applications in which intentional tampering is not an issue, magnitude DFT may be a good choice for both low- and high-processing-noise scenarios.

For robustness to "commercial quality" compression methods (better than JPEG-50 or SPIHT 1 bpp), high-GTC transforms like DCT and wavelets (subband) perform better than low-GTC transforms. Furthermore, since transforms are frequently used for image compression applications, they would leave very little room for intentional tampering without significant degradation of the image. This property would make them very suitable for watermarking applications. For other data hiding methods, with perhaps reduced robustness to intentional tampering but increased robustness to processing noise (lower-quality compression), transforms like Hadamard or Hartley would probably be more useful. For example, an average video frame is likely to suffer more processing noise than an average still-image frame. Hence, low-GTC transforms may be good choices for data hiding in video frames. Furthermore, though lower-GTC transforms are bound to have reduced robustness to intentional tampering (compared with DCT or wavelets), if the transform employed is *known*, the case is different than if the transform used is not known. There exists a high degree of freedom for the choice of the low-GTC transforms for embedding. This enhanced degree of freedom for the choice of the embedding transform could result in improved robustness to intentional tampering.

Type II and Type III (Nonlinear) Data Hiding Methods

Quantization-based data hiding methods that rely on type II and type III[1] embedding/detection principles are studied together and compared based on three key characteristics [60]:

(1) the type of the distortion reduction technique (postprocessing) employed in embedding,

(2) the form of demodulation used (detection function), and

(3) the optimization criterion utilized in determining the embedding/ detection parameters.

In the following sections, various type II and type III methods are examined and evaluated considering these three issues. The performance results of these methods, based on the above criteria, are provided in Section 5.3.

5.1 Type II Embedding and Detection

The codebook generation for type II methods is characterized by the design of $U = X + C$, which corresponds to a choice of $\alpha = 1$ within Costa's

[1]Type II can be considered a special case of type III where no postprocessing is employed.

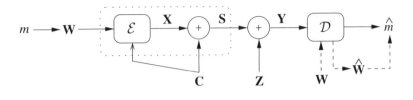

Figure 5-1   Block diagram of type II embedding and detection stages.

framework or $\mathbf{X}_t = 0$ ($\mathbf{X}_n = \mathbf{X}$) within the CAE-CID framework. The generalized channel model for type II hiding methods is displayed in Fig. 5-1. In the model, $\mathbf{W}$ is the watermark signal corresponding to the message index m to be conveyed, $\mathbf{C}$ is the host signal, $\mathbf{X}$ is the codeword, $\mathbf{S}$ is the stego signal, $\mathbf{Z}$ is the additive noise (attack), and $\mathbf{Y}$ is the distorted stego signal at the detector defined as $\mathbf{Y} = \mathbf{S} + \mathbf{Z}$. The embedder $\mathcal{E}$ imposes the power constraint as $\frac{1}{N}\|\mathbf{X}\|^2 = P$. At the detector $\mathcal{D}$ the sent message $\hat{m}$ is detected from $\mathbf{Y}$ or from an extracted estimate $\widehat{\mathbf{W}}$ of $\mathbf{W}$. Except for the codebook design, type II methods are also characterized by their $\mathcal{E}, D$ designs, which are exact inverses expressed as

$$\mathbf{S} = \mathcal{E}(\mathbf{C}, \mathbf{W}), \quad \mathbf{W} = \mathcal{D}(\mathbf{S}). \tag{5.1}$$

Chen *et al.* in [25], introduced the QIM method that outlined the codeword generation for type II methods. QIM achieves the upper bound on the hiding rate for low-level attacks (or high WNRs). In the QIM method, embedding a message into a host signal refers to quantization of the host signal by a quantizer picked from an ensemble of quantizers, where each quantizer is associated with a message letter or message index. Thus, the stego signal $\mathbf{S}$ is a quantized form of $\mathbf{C}$, and the corresponding quantization error is the codeword $\mathbf{X}$. The number of quantizers in the ensemble determines the information embedding rate. The embedding distortion is measured using a squared error distance measure, viz., $\frac{1}{N}\|\mathbf{X}\|^2 = P$, and it varies with the size and shape of the quantization cells. The orthogonality constraint, $\mathbf{X}^T\mathbf{C} = 0$, however, is relaxed by assuming that $\mathbf{C}$ is uniformly distributed over all quantization cells and the number of quantization levels is not small, such that $\mathbf{X}$ and $\mathbf{C}$ are approximately uncorrelated.

This assumption also removes the dependence of embedding and detection operations on the host signal's statistics. In practice, this can be satisfied by the *small distortions scenario*, where embedding and attack distortion powers are much less than the host signal power.

On the other hand, detection of a hidden message is achieved by the minimum distance decoder, which computes the Euclidean distances of the received signal to surrounding reconstruction points. The message index associated with the nearest reconstruction point of the corresponding quantizer is regarded as the sent message. In QIM, embedding and detection are high-dimensional operations.

A practical implementation of QIM based on dithered quantizers, viz., DM, is presented and detailed in [25] and [26]. Dithered quantizers intend to decorrelate the quantization error of a quantizer from its input [61]. In subtractive dithering, an *iid* dither vector (independent of the input) is added to the input prior to quantization and then subtracted from the quantized output. Hence, the goal (decorrelation of the quantization error) is achieved. Within the context of data hiding, the dither signal is merely a mapping from the message index, the watermark signal. Therefore, the dither signal is not genuinely random and the orthogonality between the error and the input signals is not guaranteed. In DM, each quantizer in the ensemble is generated from a base quantizer by shifting the quantization cells and reconstruction points. The stego signal is generated by quantizing the host signal with the corresponding dithered quantizer as

$$\mathbf{S} = Q_\Delta(\mathbf{C} + \mathbf{W}_m) - \mathbf{W}_m \qquad (5.2)$$

where $Q_\Delta(\cdot)$ is the high-dimensional base quantizer with reconstruction points Δ apart, and $\mathbf{W}_m$ is the watermark signal corresponding to the message indexed by m, $1 \leq m \leq M$, where each component W_{m_i}, $1 \leq i \leq N$, of $\mathbf{W}_m$ is a representation from a set $\Omega \in \mathfrak{R}$. Consequently, the codeword $\mathbf{X}$ is defined as

$$\mathbf{X} = (Q_\Delta(\mathbf{C} + \mathbf{W}_m) - \mathbf{W}_m) - \mathbf{C}. \qquad (5.3)$$

The power constraint on the embedding distortion $\mathbf{X}$ is controlled by adjusting the quantization step size Δ.

For the sake of practicality, $Q_\Delta(\cdot)$ can be considered to be a product quantizer generated by a Cartesian product of N uniform scalar quantizers, $q_\Delta(\cdot)$, each with step size Δ such that

$$q_\Delta(C) = i\Delta, \quad \text{for } i\Delta - \frac{\Delta}{2} \le C < i\Delta + \frac{\Delta}{2}. \tag{5.4}$$

Therefore, embedding can be viewed as N successive scalar quantizations, of the coefficients of $\mathbf{C} = (C_1, \ldots, C_N)$, dithered with the watermark signal vector $\mathbf{W}_m = (W_{m_1}, \ldots, W_{m_N})$. Each distinct component of the watermark (dither) signal is associated with a quantizer that is generated by properly shifting the reconstruction points of $q_\Delta(\cdot)$. The amount of shifting is determined by the number of possible values a watermark sample can take (the number of quantizers). For maximum separation of the reconstruction points of embedding quantizers, the watermark sample values are equally spaced along an interval of length that is equal to quantization step size Δ, i.e., $[-\Delta/2, \Delta/2)$. It should be noted that since the watermark signal is the subtractive dither signal, the sample values represented by the form $W_m + i\Delta$ for $i \in \mathcal{Z}$, where $\mathcal{Z}$ is the set of all integers, lead to the same dithered quantizer. (In other words, shifts differing by an integer multiple of Δ correspond to the same quantizer.) Considering a d-ary watermark sample, the set Ω that contains the d possible sample values is defined as

$$\Omega = \left\{ \delta + i\Delta, \delta + \frac{\Delta}{d} + i\Delta, \delta + 2\frac{\Delta}{d} + i\Delta, \ldots, \delta + (d-1)\frac{\Delta}{d} + i\Delta \right\} \tag{5.5}$$

where δ is a uniform random variable in $[-\frac{\Delta}{2}, \frac{\Delta}{2})$ and $i \in \mathcal{Z}$. As a result, reconstruction points and quantization cells of each quantizer in the ensemble are shifted by $\frac{\Delta}{d}$ with respect to each other. The reconstruction points of the embedding quantizers are also known to the detector for the extraction of the sent message. At the detector, the hidden message is extracted by the minimum distance decoder as

$$\hat{m} = \arg\min_m \| \mathbf{Y} - (q_\Delta(\mathbf{Y} + \mathbf{W}_m) - \mathbf{W}_m) \|. \tag{5.6}$$

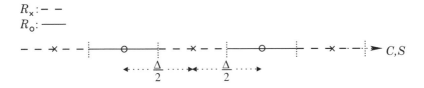

Figure 5-2   Reconstruction points of dithered quantizers corresponding to a binary watermark (dither) signal.

Figure 5-2 displays the reconstruction points of the dithered quantizers associated with the two watermark samples. The reconstruction points of the two quantizers are $\frac{\Delta}{2}$ apart. The decision regions denoted by $R_\times$ and $R_\circ$ determine the sustainable amount of noise for successful extraction of the message. The stego signal **S** is generated by quantizing each host signal coefficient C with the quantizer pointed by the binary watermark sample W of **W** to be embedded. (Accordingly, embedding of the watermark sample associated with $\times$ or $\circ$ refers to translation of the host signal coefficient C in the direction of the nearest $\times$ or $\circ$.) Similarly, detection of a sent message is achieved by determining the nearest reconstruction points, denoted $\times$ or $\circ$, to the coefficients of the received signal **Y**.

The main disadvantage of type II methods is that they perform well only if the attack is not severe (less than distortion P). In other words, their performance is equivalent to that of optimal design *only for the low attack case* (see Section 3.1). For all other attack levels, there is a performance gap with the upper bound, which increases with the attack level. This is due to the nonoptimal codebook design based on $\alpha = 1$ or equivalently $\mathbf{X}_t = \mathbf{0}$, which undermines the dependency of codebook generation to the channel noise level. The poor performance of type II methods with increasing attack levels is improved by the modifications proposed by the class of methods called type III.

5.2   Type III Embedding and Detection Methods

The data hiding rate (payload) vs robustness performance of type II methods is substantially improved by enhancing the functionality of the embedder

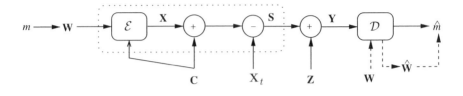

Figure 5-3   Block diagram of type III embedding and detection stages.

with further processing capabilities (i.e., thresholding, distortion compensation, Gaussian mapping) [9], [21], [22], [36]. In type III methods, embedding quantization is followed by a processing stage (postprocessing) that generates the stego signal. The improvement in the performance of type III methods, compared with type II, at the same noise level can be explained by the fact that codebook design depends on channel noise level or by the deviation from the nonoptimal design of $\mathbf{X}_t = \mathbf{0}$ through the added processing. Alternately, in terms of Costa's framework, the improvement can be attributed to the effective value of α used in codebook generation, which is less than 1 rather than being equal to 1, as the latter is optimal for the no-attack case. Data hiding methods with postprocessing abilities enable the embedder to increase the distance between the reconstruction points of quantizers at a fixed embedding distortion. Therefore, they have improved detection capabilities for any finite WNR level (type II is optimal only for the case of infinite WNR). On the other hand, since the detector is blind to the additional processing at the embedder, its structure is not altered.

The channel model for type III hiding methods, based on the model for type II methods given in Fig. 5-1, is displayed in Fig. 5-3. In the model, $\mathbf{X}$ is the type II codeword (embedding distortion introduced due to the quantization), $\mathbf{X}_t$ is the processing distortion, and the channel output is $\mathbf{Y} = \mathbf{C} + \mathbf{X} - \mathbf{X}_t + \mathbf{Z}$. The processing distortion $\mathbf{X}_t$ is derived from $\mathbf{X}$ by postprocessing depending on the expected noise level. The type III codeword that yields the stego signal, $\mathbf{S} = \mathbf{C} + \mathbf{X}_n$, is defined as $\mathbf{X}_n = \mathbf{X} - \mathbf{X}_t$. Correspondingly, the embedder imposes the power constraint as $\frac{1}{N}\|\mathbf{X}_n\|^2 = P$.

In type III methods, since the detector is not aware of the processing at the embedder, the processing distortion $\mathbf{X}_t$ can effectively be considered to be a part of the channel noise at the detector. Therefore, the type II codeword $\mathbf{X}$, which would yield an errorless extraction of the watermark signal $\mathbf{W}$, is distorted by two sources of noise, viz., the attack $\mathbf{Z}$ and the processing distortion $\mathbf{X}_t$. (In other words, the signal $\mathbf{C} + \mathbf{X}$ refers to a signal quantized by the quantizer(s) associated with the watermark signal $\mathbf{W}$, and the embedded $\mathbf{W}$ can be perfectly recovered from this signal.) Therefore, the effective noise at the detector that distorts the embedded watermark signal is represented as $\mathbf{Z}_{eff} = \mathbf{Z} - \mathbf{X}_t$. In type III methods, the invertibility condition on the $\mathcal{E}, D$ pair is sacrificed as a result of the processing that follows quantization of the host signal, $\mathcal{D}(\mathcal{E}(C, W)) \neq W$.

Performance of type III hiding methods vary based on three factors: the type of postprocessing that is incorporated with type II embedding, the choice of demodulation function used in message extraction, and the criterion used for optimizing the embedding and detection parameters. Therefore, the performance of any type III data hiding method can be evaluated further by considering these three issues.

5.2.1 Postprocessing Types

There are three types of postprocessing employed in type III embedder/detector designs. These are:

- distortion compensation,
- thresholding, and
- Gaussian mapping.

In [9], Chen *et al.* identified the capacity-achieving variant of QIM as DC-QIM (see Chapter 2.4). In DC-QIM, the quantization index modulated signal is perturbated by subtracting the $1 - \alpha^*$ scaled version of the embedding distortion $\mathbf{X}$. Therefore, $\mathbf{X}_t = (1 - \alpha^*)\mathbf{X}$, $\rho = 1$, and $\mathbf{X}_n = \alpha^*\mathbf{X}$. Ramkumar *et al.* [21] proposed a thresholding type of postprocessing in which the magnitude of distortions that can be introduced to the host signal samples are limited to $\pm\frac{\beta}{2}$. Hence, the type III codeword $\mathbf{X}_n$ is generated by limiting the values of $\mathbf{X}$, $\mathbf{X}_n = \min(|\mathbf{X}|, \frac{\beta}{2})\text{sign}(\mathbf{X})$.

TABLE 5-1
Expressions for $\mathbf{X}_t$ and $\mathbf{X}_n$

PROCESSING, $\mathcal{P}$	PROCESSING DISTORTION, $\mathbf{X}_t$	CODEWORD, $\mathbf{X}_n$
Thresholding	$\max(0, |\mathbf{X}| - \frac{\beta}{2})\text{sign}(\mathbf{X})$	$\min(|\mathbf{X}|, \frac{\beta}{2})\text{sign}(\mathbf{X})$
Distortion compensation	$(1 - \alpha)\mathbf{X}$	$\alpha\mathbf{X}$
Gaussian mapping	$-\sigma_v Q^{-1}\left(\frac{\mathbf{X}+\frac{\Delta}{2}}{\Delta}\right)$	$\mathbf{X} - \mathbf{X}_t$

The processing distortion $\mathbf{X}_t$, in this case, is the thresholding noise, $\mathbf{X}_t = \max(0, |\mathbf{X}| - \frac{\beta}{2})\text{sign}(\mathbf{X})$. Perez-Gonzalez *et al.* [36], considering uniform scalar quantization, proposed to generate the processing distortion $\mathbf{X}_t$ from $\mathbf{X}$ by transforming each *iid* component X into a zero-mean Gaussian distributed random variable with a variance of σ_v^2, $\mathbf{X}_t = -\sigma_v Q^{-1}\left(\frac{\mathbf{X}+\frac{\Delta}{2}}{\Delta}\right)$, where $Q^{-1}(\cdot)$ is the inverse Gaussian Q-function.

In type III methods, the parameters α, β, and σ_v, depending on the type of postprocessing, are selected in such a way that the power constraint $\frac{1}{N}\|\mathbf{X}_n\|^2 = P$ is satisfied and the performance at the presumed noise (attack) level is maximized. Corresponding expressions for the processing distortion $\mathbf{X}_t$ and the codeword $\mathbf{X}_n$ for the three types of postprocessing are as given in Table 5-1.

5.2.1.1 Vectoral Embedding and Detection

The optimal processing, within the CAE-CID framework, requires that the processing distortion $\mathbf{X}_t$ be a linear function of the processing distortion $\mathbf{X}$. Accordingly, the power σ_X^2 of the embedding distortion $\mathbf{X}$ corresponding to the distortion-compensation type of postprocessing can be computed in the limit, using $\frac{1}{N}\|\mathbf{X}_n\|^2 = P$, as

$$\sigma_X^2 = \frac{1}{N}\|\mathbf{X}\|^2 = \frac{1}{N}\left\|\frac{\mathbf{X}_n}{\alpha^*}\right\|^2 = \frac{(P + \sigma_Z^2)^2}{P} \tag{5.7}$$

where $\alpha^* = \frac{P}{P+\sigma_Z^2}$. It should be noted that the variance of the *iid* components of the channel input $\mathbf{X}$ (the power of the input $\mathbf{X}$) in Eq. (3.15) is the same as the power of the optimal embedding distortion $\mathbf{X}$ found in Eq. (5.7), $\sigma_X = \sigma_X^*$. Therefore, distortion compensation is the optimal post-processing when the embedding distortion is Gaussian distributed. This can be satisfied by the use of high-dimensional quantization for embedding which yields a Gaussian distributed quantization error. However, a capacity-achieving embedding/detection scheme based on thresholding or Gaussian mapping types of postprocessing is not possible because the relation between $\mathbf{X}$ and $\mathbf{X_t}$ is not linear.

5.2.1.2 Scalar Embedding and Detection

In some practical cases, scalar quantization rather than high-dimensional vector quantization is employed at the embedder. $\mathbf{X}$ is an *iid* vector with a non-Gaussian distribution. Therefore, the optimal postprocessing is not necessarily distortion compensation. For the scalar quantization case, the embedding operation of all embedding/detection techniques can be represented by a form of dithered quantization. Thus, each component X of the embedding distortion $\mathbf{X}$, defined as $\mathbf{X} = q_\Delta(\mathbf{C}, \mathbf{W}_m) - \mathbf{W}_m - \mathbf{C}$, is uniformly distributed. However, the processing distortion $\mathbf{X}_t$ and its dependency on $\mathbf{X}$ are different for the three types of postprocessing.

Eggers *et al.*, in [22] optimized the value of α for scalar quantization, rather than assuming $\alpha^* = \frac{P}{P+\sigma_Z^2}$, and provided the approximation

$$\alpha = \sqrt{\frac{P}{P + 2.71\sigma_Z^2}}. \tag{5.8}$$

Expressions for the optimal values of Δ and the threshold β based on the expected attack level were reported in [21]. Although [36] does not provide the optimal σ_v values for Gaussian mapping, the optimization procedure is straightforward.

5.2.2 Forms of Demodulation

Detection of the sent message is achieved either by sample-wise hard decisions or soft decisions based on the availability of the set of watermark signals at the extractor side. The presence of watermark signals leads to an improved detection of the sent message because they can be utilized in detection operation [21], [26].

There are two forms of demodulation employed in detection of the sent message. In [22], [26], [36], demodulation of the sent message, from the received signal $\mathbf{Y}$, is realized by minimum distance decoding, and in [21], demodulation takes the form of a maximum correlation rule.

5.2.2.1   Minimum Distance Decoding

With the use of minimum distance decoding, detection is simply the quantization of the received signal $\mathbf{Y}$ by all quantizers in the ensemble. The message letter or message index associated with the quantizer that yields the minimum Euclidean distance to received $\mathbf{Y}$ is deemed to be the sent message. The general form of minimum distance decoding based on dithered quantization can be rewritten, in terms of $\mathbf{Y}_m = \mathbf{Y} + \mathbf{W}_m$, as

$$\hat{m} = \mathcal{D}(\mathbf{Y}) = \arg\min_{m} \|\mathbf{Y}_m - Q_\Delta(\mathbf{Y}_m)\|, \quad 1 \le m \le M. \qquad (5.9)$$

It should be noted that Eq. (5.9) is a minimization of the quantization error over all quantizers. For the case of scalar quantization, $Q_\Delta(\cdot)$ takes the form of dithered quantizer $q_\Delta(\cdot)$, Eq. (5.6).

Figure 5-4 displays the detectors for the binary signaling case where the embedding operation is based on scalar quantization. In the figure, the symbols $\times$ and $\circ$ denote the reconstruction points of the quantizers associated with the watermark sample values of $-\frac{\Delta}{4}$ and $\frac{\Delta}{4}$. (However, it should be noted that within the scope of DM, any two sample values with $\frac{\Delta}{2}$ difference are valid choices; see Eq. (5.5).)

When the extractor has no access to the watermark signals but knows only the reconstruction points, each sample of the embedded watermark signal is detected from each coefficient Y of the received signal $\mathbf{Y}$ by individual hard decisions as

$$\widehat{W}_i = \arg\min_{W_i \in \Omega} \|Y_i + W_i - q_\Delta(Y_i + W_i)\| \quad \text{for } i = 1, \dots, N \qquad (5.10)$$

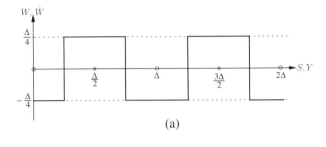

(a)

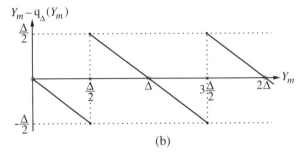

(b)

Figure 5-4   Demodulation for DM based on (a) hard decisions and (b) soft decisions.

where Ω is the set of signal representations for watermark samples. Equation (5.10) is based on determining the minimum Euclidean distance of the received signal coefficients to reconstruction points that can equivalently be achieved by mapping each coefficient Y over the square wave function displayed in Fig. 5-4a. Then, the extracted binary watermark samples, $\widehat{W}_1, \ldots, \widehat{W}_N$, are combined into the sequence $\widehat{\mathbf{W}}$ to generate the embedded watermark signal.

On the other hand, when the watermark signals are present at the detector, detection of each sample is by soft decisions. Accordingly, each coefficient Y_m of the signal $\mathbf{Y}_m$ that is obtained from the received signal $\mathbf{Y}$ is mapped over the sawtooth function displayed in Fig. 5-4b. The norm of the resulting signal values is the distance between $\mathbf{Y}$ and $\mathbf{W}_m$. Hence, the watermark signal that has the minimum distance to $\mathbf{Y}$ is regarded as the embedded signal.

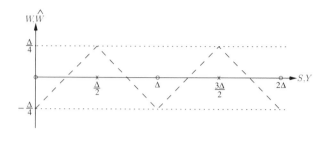

Figure 5-5 Periodic extraction function corresponding to soft decisions.

5.2.2.2 Maximum Correlation Rule

When the demodulation scheme is based on a maximum correlation detector, watermark signals are assumed to be present at the detector. In this form of demodulation, at first an estimate $\widehat{\mathbf{W}}$ of the embedded watermark signal is extracted from the received signal by soft decisions. Then, the sent message is detected by matching the estimate of the embedded watermark signal to one of the watermark signals using a correlation-based similarity measure as

$$\widehat{\mathbf{W}} = \mathcal{D}(\mathbf{Y}),$$

$$\hat{m} = \arg\max_{m} \frac{\mathbf{W}_m \widehat{\mathbf{W}}}{\|\mathbf{W}_m\| \, \|\widehat{\mathbf{W}}\|}, \quad 1 \leq m \leq M. \tag{5.11}$$

Since the hard decisions are caused by the discontinuities in the extraction function, Fig. 5-4a, [21] has proposed a continuous periodic triangular extraction function. Figure 5-5 displays the corresponding function used for extracting embedded binary watermark samples that are confined to values $-\frac{\Delta}{4}$ and $\frac{\Delta}{4}$ for maximum separation, $\Omega = \{-\frac{\Delta}{4}, \frac{\Delta}{4}\}$. An estimate of the embedded watermark signal is obtained by mapping each coefficient of $\mathbf{Y}$ over the periodic triangular function, rather than making a hard decision by the Euclidean distance decoder. As a result, each extracted sample $\widehat{W}$ is a real-valued signal in the range of $[-\frac{\Delta}{4}, \frac{\Delta}{4}]$. Message detection is achieved by combining the sample estimates into $\widehat{\mathbf{W}} = (\widehat{W}_1, \ldots, \widehat{W}_N)$ and then matching $\widehat{\mathbf{W}}$ to one of $\mathbf{W}_1, \ldots, \mathbf{W}_M$.

5.2.3 Optimization Criteria for Embedding and Detection Parameters

Embedding and detection operations are controlled by a pair of parameters. The values for these parameters are optimized for the given channel noise and permitted distortion levels, σ_Z^2 and P.

One of the parameters common to all techniques is Δ, which designates the distance between the reconstruction points of the embedding quantizers. The choice of Δ determines the embedding distortion due to quantization, and it is known to both embedder and detector. The other parameter controls the amount of processing distortion introduced into the quantized signal (type II embedded signal) by postprocessing and limits the distortion due to the embedding operation to the permitted amount. This parameter is known only to the embedder and parameterized as β, α, or σ_V depending on the type of postprocessing. The values for the two interdependent parameters can be optimized based on various performance criteria, as discussed in the following sections.

5.2.3.1 Optimization of Parameters for Vectoral Embedding and Detection

In [9], researchers optimized the embedding/detection parameters by maximizing the ratio of the embedding distortion to the sum of processing and channel distortions, $\left(\frac{\sigma_X^2}{\sigma_{X_t}^2 + \sigma_Z^2} \right)$, at the extractor as

$$(\Delta, \sigma_{X_t}^2) = \arg \max_{\Delta, \sigma_{X_t}^2} \left\{ \frac{\sigma_X^2}{\sigma_{X_t}^2 + \sigma_Z^2} \;\middle|\; \sigma_Z^2, \frac{1}{N}\|\mathbf{X}_n\|^2 = P, \mathbf{X_t} \right\}. \tag{5.12}$$

With the use of high-dimensional quantization for embedding and detection, the marginal pdf of embedding distortion $\mathbf{X}$ approximates Gaussian distribution, and consequently, distortion compensation becomes the optimal postprocessing. Hence, for the given channel noise level, Δ and α are selected in such a way that Eq. (5.12) is satisfied, where $\mathbf{X}_t = (1 - \alpha)\mathbf{X}$ and $\mathbf{X}_n = \alpha\mathbf{X}$, i.e., $\sigma_{X_t}^2 = (1 - \alpha)^2 \sigma_X^2$ and $\sigma_X^2 = \frac{P}{\alpha^2}$. This leads to $\alpha = \frac{P}{P + \sigma_Z^2}$, which is in accord with the results of Section 3.1 due to the duality between the two channel models.

5.2.3.2   Optimization of Parameters for Scalar Embedding and Detection

Researchers [21], [22], [36] have modeled the effective noise that distorts the embedded watermark signal in terms of the statistics of the channel noise $\mathbf{Z}$ and the processing distortion $\mathbf{X}_t$, $\mathbf{Z}_{eff} = \mathbf{Z} - \mathbf{X}_t$. The optimum values for embedding/detection parameters are then selected in such a way that the distortion in the extracted watermark signal is minimized.

When the host signal is uniformly distributed over all quantization intervals, the embedding distortion X introduced into each host signal coefficient C is uniformly distributed in $[-\frac{\Delta}{2}, \frac{\Delta}{2}]$. For the thresholding type of postprocessing, the parameters are the step size Δ and the threshold β. The corresponding pdf and statistics of processing distortion X_t and the codeword X_n are expressed as

$$f_{X_t}(x_t) = \frac{\beta}{\Delta}\delta(x_t) + \frac{1}{\Delta}\text{rect}(\Delta - \beta), \tag{5.13}$$

$$m_{X_t} = 0, \tag{5.14}$$

$$\sigma_{X_t}^2 = \frac{(\Delta - \beta)^3}{12\Delta}, \tag{5.15}$$

$$f_{X_n}(x) = \frac{1}{\Delta}\text{rect}(\beta) + \frac{\Delta - \beta}{2\Delta}\left(\delta(x_n - \frac{\beta}{2}) + \delta(x_n + \frac{\beta}{2})\right), \tag{5.16}$$

$$m_{X_n} = 0, \tag{5.17}$$

$$\sigma_{X_n}^2 = \frac{\beta^2}{12\Delta}(3\Delta - 2\beta) \tag{5.18}$$

where $\text{rect}(x)$ is the rectangular function in x with a value of 1 in the interval $(-\frac{1}{2}, \frac{1}{2})$ and zero elsewhere. Similarly, for the distortion-compensation type of postprocessing, corresponding pdf's and statistics are found in terms of Δ and α as

$$f_{X_t}(x_t) = \frac{1}{(1 - \alpha)\Delta}\text{rect}((1 - \alpha)\Delta), \tag{5.19}$$

$$m_{X_t} = 0, \tag{5.20}$$

$$\sigma_{X_t}^2 = \frac{(1 - \alpha)^2 \Delta^2}{12}, \tag{5.21}$$

$$f_{X_n}(x) = \frac{1}{\alpha \Delta} \text{rect}(\alpha \Delta), \tag{5.22}$$

$$m_{X_n} = 0, \tag{5.23}$$

$$\sigma_{X_n}^2 = \frac{\alpha^2 \Delta^2}{12}. \tag{5.24}$$

When postprocessing takes the form of Gaussian mapping, X_t is a zero-mean Gaussian random variable with variance σ_v^2 and the parameters are Δ and σ_v. However, as the dependency between X and X_t is through a Gaussian transformation, the pdf of X_n is not a straightforward one, but its statistics can be calculated as

$$E[X_n^k] = \int_{-\frac{\Delta}{2}}^{\frac{\Delta}{2}} \left(x + \sigma_v Q^{-1} \left(\frac{x + \frac{\Delta}{2}}{\Delta} \right) \right)^k \frac{1}{\Delta} dx. \tag{5.25}$$

Figures 5-6 and 5-7 display $f_X(x)$, $f_{X_t}(x_t)$, and $f_{X_n}(x_n)$ for thresholding and distortion-compensation types of postprocessing, respectively.

Given that the host signal is *iid*, $\mathbf{X}$ and $\mathbf{X}_t$ are *iid* random vectors with the marginal distributions given as previously, since the embedding operation is memoryless. It should also be noted that for large N, the distortion P introduced into host signal $\mathbf{C}$, due to the embedding operation, is equal to $\sigma_{X_n}^2$, i.e., $\frac{1}{N} \|\mathbf{X}_n^2\| = P$.

Assuming that Z and X_t are independent, the resulting pdf of Z_{eff}, $f_{Z_{eff}}(z_{eff})$, can be computed by the convolution of the individual pdf's $f_Z(z)$

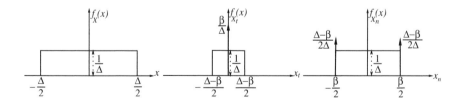

Figure 5-6   Probability density functions (left) $f_X(x)$, (center) $f_{X_t}(x_t)$, and (right) $f_{X_n}(x_n)$ corresponding to thresholding type of processing for $0 < \beta < \Delta$.

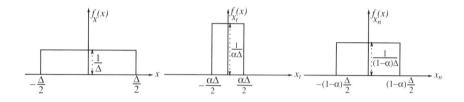

Figure 5-7   Probability density functions (left) $f_X(x)$, (center) $f_{X_t}(x_t)$, and (right) $f_{X_n}(x_n)$ corresponding to distortion-compensation type of processing for $\alpha < 1$.

and $f_{X_t}(x_t)$ as

$$f_{Z_{eff}}(z_{eff}) = \int_{-\infty}^{\infty} f_Z(z_{eff} - x)f_{X_t}(x)dx. \tag{5.26}$$

Thus, for $Z \sim \mathcal{N}(0, \sigma_Z^2)$, $f_{Z_{eff}}(z_{eff})$ corresponding to the thresholding type of postprocessing is derived as

$$f_{Z_{eff}}(z_{eff}) = \frac{\beta}{\Delta\sqrt{2\pi\sigma_Z^2}}\exp\left(-\frac{z_{eff}^2}{2\sigma_Z^2}\right)$$
$$+ \frac{1}{2\Delta}\left(\mathrm{erf}\left(\frac{z_{eff} + \frac{\Delta-\beta}{2}}{\sqrt{2}\sigma_Z}\right) - \mathrm{erf}\left(\frac{z_{eff} - \frac{\Delta-\beta}{2}}{\sqrt{2}\sigma_Z}\right)\right) \tag{5.27}$$

where $\mathrm{erf}(\cdot)$ is the Gaussian error function, $\mathrm{erf}(z) = \frac{2}{\pi}\int_0^z \exp^{-x^2} dx$. Similarly, in distortion-compensation and Gaussian mapping cases, respectively, $f_{Z_{eff}}(z_{eff})$ is expressed as

$$f_{Z_{eff}}(z_{eff}) = \frac{1}{2(1-\alpha)\Delta}\left(\mathrm{erf}\left(\frac{z_{eff} + \frac{(1-\alpha)\Delta}{2}}{\sqrt{2}\sigma_Z}\right) - \mathrm{erf}\left(\frac{z_{eff} - \frac{(1-\alpha)\Delta}{2}}{\sqrt{2}\sigma_Z}\right)\right) \tag{5.28}$$

and

$$f_{Z_{eff}}(z_{eff}) = \frac{1}{\sqrt{2\pi(\sigma_Z^2 + \sigma_v^2)}}\exp\left(-\frac{z_{eff}^2}{2(\sigma_Z^2 + \sigma_v^2)}\right). \tag{5.29}$$

The embedding-detection parameters are optimized by proper selection of the step size Δ and the amount of processing distortion $\sigma_{X_t}^2$. Such a selection can be based on one of the three criteria for the given statistics of $\mathbf{Z}_{eff}$.

5.2.3.3 Maximizing Correlation

With this criterion, the selection of parameters is based on maximizing the normalized correlation between the embedded and the extracted watermark signals [21]. Since $\mathbf{Z}_{eff}$ is the noise that distorts the type II codeword $\mathbf{X}$ corresponding to watermark signal $\mathbf{W}$, the signal $\widehat{\mathbf{W}}$ extracted from $\mathbf{Y}$ can be expressed in terms of $\mathbf{Z}_{eff}$ and $\mathbf{W}$ using the extraction function shown in Fig. 5-5. (Note that if $\mathbf{Z}_{eff} = 0$, then $\mathbf{W}=\widehat{\mathbf{W}}$.) Hence, a binary distributed watermark signal sample W with values in $\{-\frac{\Delta}{4}, \frac{\Delta}{4}\}$ embedded in a host signal coefficient is extracted as

$$\widehat{W} = \begin{cases} (\dfrac{(2i+1)\Delta}{4} - Z_{eff})(-1)^i, \ i\dfrac{\Delta}{2} < Z_{eff} \le \dfrac{(i+1)\Delta}{2} \ , \ \ i \in \mathcal{Z} \text{ if } W = \dfrac{\Delta}{4}, \\ (-\dfrac{(2i+1)\Delta}{4} + Z_{eff})(-1)^i, \ i\dfrac{\Delta}{2} < Z_{eff} \le \dfrac{(i+1)\Delta}{2} \ , \ \ i \in \mathcal{Z} \text{ if } W = -\dfrac{\Delta}{4}. \end{cases}$$

$$(5.30)$$

Due to memoryless embedding/detection and attack schemes, the vectors $\mathbf{W}$ and $\widehat{\mathbf{W}}$ are *iid* with sample values W and $\widehat{W}$. Hence the normalized correlation ρ between $\mathbf{W}$ and $\widehat{\mathbf{W}}$ can be analytically computed for large N as

$$\begin{aligned} \rho &= E\left[\frac{\mathbf{W}^T\widehat{\mathbf{W}}}{\|\mathbf{W}\| \ \|\widehat{\mathbf{W}}\|} \right] \\ &= \frac{E[W\widehat{W}]}{\sqrt{E[W^2]E[\widehat{W}^2]}} \\ &= \frac{R(1)}{\sqrt{R(2)}} \end{aligned}$$

$$(5.31)$$

where $E[W\widehat{W}]$ is the first joint moment of the random variables W and $\widehat{W}$, and

$$R(p) = 2 \sum_{i=0}^{i=\infty} \int_{\frac{i\Delta}{2}}^{\frac{(i+1)\Delta}{2}} \left(\left(\frac{(2i+1)\Delta}{4} - z_{\textit{eff}} \right) (-1)^i \right)^p f_{Z_{\textit{eff}}}(z_{\textit{eff}}) dz_{\textit{eff}}.$$

$$(5.32)$$

Therefore, the optimal parameter values for the utilized postprocessing technique is computed by maximizing Eq. (5.31) over Δ and $\sigma_{X_t}^2$ using the pdf's given in Eqs. (5.27)–(5.29) for the given channel noise level and permitted distortion as

$$(\Delta, \sigma_{X_t}^2) = \arg\max_{\Delta, \sigma_{X_t}^2} \left\{ \rho \,\middle|\, \sigma_Z^2, \mathbf{X}_t \in \mathcal{X}_t, \sigma_{X_n}^2 = P \right\} \qquad (5.33)$$

where

$$\mathcal{X}_t = \left\{ \max\left(\mathbf{0}, |\mathbf{X}| - \frac{\beta}{2}\right) \text{sign}(\mathbf{X}), (1-\alpha)\mathbf{X}, -\sigma_v Q^{-1}\left(\frac{\mathbf{X} + \frac{\Delta}{2}}{\Delta}\right) \right\}$$

$$(5.34)$$

and $\mathbf{X} = q_\Delta(\mathbf{C} + \mathbf{W}) - \mathbf{W} - \mathbf{C}$.

5.2.3.4   Minimizing Probability of Error

The embedding/detection parameters are selected to minimize the probability of error in detecting an embedded watermark sample [36]. Since $Z_{\textit{eff}}$ indicates the deviation of the received signal coefficient Y from the reconstruction points, the probability of detection error, P_e, can be calculated by integrating $f_{Z_{\textit{eff}}}(z_{\textit{eff}})$ over all decision regions but excluding the one associated with the sent sample as

$$P_e = P\{Y \notin \mathcal{R}_W \mid W\} \qquad (5.35)$$

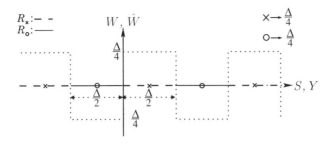

Figure 5-8 Embedding and detection of a binary watermark sample.

where $\mathcal{R}_W$ denotes the decision region associated with the sample W. For the binary signaling case depicted in Fig. 5-8, the symbols $\times$ and $\circ$ denote the reconstruction points of two quantizers associated with sample values $\frac{\Delta}{4}$ and $-\frac{\Delta}{4}$, respectively. The decision regions $R_\times$ and $R_\circ$ are used to map the received signal coefficient Y to $\frac{\Delta}{4}$ or $-\frac{\Delta}{4}$ by hard decisions. Assuming $\frac{\Delta}{4}$ and $-\frac{\Delta}{4}$ are equally likely to be embedded, the corresponding P_e is calculated as

$$
P_e = P\left\{ \left\| Y + \frac{\Delta}{4} - q_\Delta\left(Y + \frac{\Delta}{4}\right)\right\| > \left\| Y - \frac{\Delta}{4} - q_\Delta\left(Y - \frac{\Delta}{4}\right)\right\| \,\middle|\, w = \frac{\Delta}{4} \right\},
$$

$$
= P\left\{ Y \in \mathcal{R}_\circ \,\middle|\, w = \frac{\Delta}{4} \right\},
$$

$$
= \int_{\mathcal{R}_\circ} f_{Z_{eff}}\left(z_{eff} - \frac{\Delta}{4}\right) dz_{eff}. \tag{5.36}
$$

Then, the parameters can be selected to minimize P_e for the given P, σ_Z^2, and the type of postprocessing as

$$
(\Delta, \sigma_{X_t}^2) = \arg \min_{\Delta, \sigma_{X_t}^2} \left\{ P_e \,\middle|\, \sigma_Z^2, X_t \in \mathcal{X}_t, \sigma_{X_n}^2 = P \right\} \tag{5.37}
$$

where $\mathcal{X}_t$ is given in Eq. (5.34).

5.2.3.5   Maximizing Mutual Information

The parameters are selected to maximize the mutual information between the embedded watermark sample W and the received signal coefficient Y [22]. The mutual information between W and Y is expressed as

$$I(W, Y) = H(Y) - H(Y|W) \tag{5.38}$$

where $H(\cdot)$ is the differential entropy of a random variable in bits that is defined as $H(X) = -\int_{-\infty}^{\infty} f_X(x) \log_2[f_X(x)] \, dx$. As the erroneous detection of W from Y is due to the noise Z_{eff}, $H(Y|W)$ in Eq. (5.38) can be computed in terms of the effective noise pdf conditioned on W, $f_{Z_{eff}|W}(z_{eff}|w)$. The pdf $f_{Z_{eff}|W}(z_{eff}|w)$ can be calculated over any quantization interval Δ, because the signal constellation is periodic with Δ (reconstruction points corresponding to the quantizer associated with W are Δ apart). However, one should take into account that when Z_{eff} is heavy tailed (the range of $f_{Z_{eff}}(z_{eff})$ is larger than Δ), its pdf will be wrapped around Δ due to the periodicity. Consequently, $H(Y)$ is computed from $H(Y|W)$ by averaging it over W. (Assuming that all samples $W \in \Omega$ are equally likely, $H(Y)$ is obtained as $\frac{1}{|\Omega|} \sum_{W \in \Omega} H(Y|W)$.) With this criterion, optimization of parameter values is by maximizing Eq. (5.38) for the given constraints over Δ and $\sigma_{X_t}^2$ as

$$(\Delta, \sigma_{X_t}^2) = \arg \max_{\Delta, \sigma_{X_t}^2} \left\{ I(Y, W) \, \middle| \, \sigma_Z^2, X_t \in \mathcal{X}_t, \sigma_{X_n}^2 = P \right\}. \tag{5.39}$$

The use of Eq. (5.38) also enables computation of the maximum hiding rate in bits per host signal coefficient achievable with a particular embedding/detection technique. Therefore, it is a useful performance evaluation tool.

5.3   Performance Comparisons

Figure 5-9 displays the achievable data hiding rates of various embedding/detection techniques for the binary signaling case, obtained using

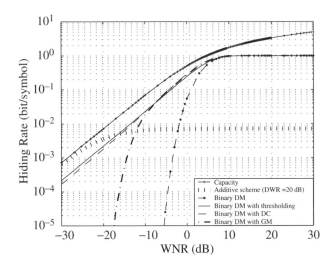

Figure 5-9 Comparison of the hiding rates corresponding to various hiding methods considering binary signaling obtained for $P = 10$.

Eqs. (3.4) and (5.38), compared with hiding rates of type I (additive scheme) and optimal type III (capacity). The embedding/detection parameters for type II and type III methods are selected so that the hiding rate is maximized, Eq. (5.39). The additive scheme (type I) and DM (type II) have preferable performances, respectively, at very low and very high WNRs. For DM, the gap with the upper bound at higher WNRs is due to binary signaling. Thus, the performance can be improved for multi-level signal representations. The poor performance of both methods in mid-WNR range is due to nonoptimal codebook designs, as discussed in Section 3.4. In the former, the codebook design does not utilize the host signal, and in the latter, the design disregards the channel noise level.

The type III versions of DM, implemented by incorporating the embedding of DM with thresholding, distortion-compensation, and Gaussian-mapping types of postprocessing, have better performances than DM due to the deviation from the optimistic "low-noise" assumption in the codebook design. These methods have significantly improved performances in the mid-WNR range; however, in order to achieve higher rates, embedding through scalar quantization has to be substituted by high-dimensional vector quantization.

Type III methods employing thresholding and distortion compensation types of postprocessing perform closely in the whole WNR range. On the other hand, Gaussian-mapping processing has a comparable performance only for WNRs higher than -7.8 dB. Below that range the rate drops rapidly. At WNRs lower than -8.7 dB, thresholding performs marginally better, while from -8.7 dB to -7 dB, distortion compensation performs best. Above -7 dB, both distortion compensation and Gaussian mapping are the preferred postprocessing types. Figures 5-10–5-13 show the hiding rates for the corresponding methods with multilevel signaling. With the decreasing noise level and higher signal representation levels, all methods yield similar data hiding rates as the need for postprocessing is reduced. Ultimately, when there is no noise, DM is the optimal embedding/detection technique.

The normalized-correlation ρ and probability-of-error P_e performances for the considered methods are given, respectively, in

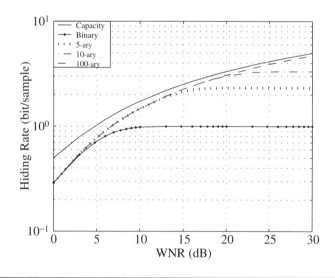

Figure 5-10   Data hiding rates for DM with binary, 5-ary, 10-ary, and 100-ary signaling.

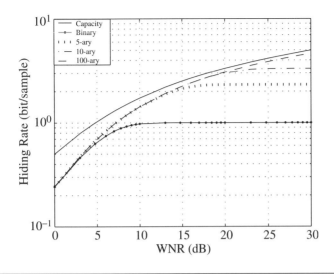

Figure 5-11 Data hiding rates for DM followed by thresholding type of postprocessing with binary, 5-ary, 10-ary, and 100-ary signaling.

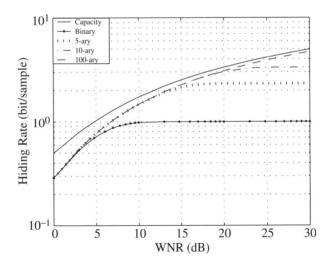

Figure 5-12 Data hiding rates for DM followed by distortion-compensation type of postprocessing with binary, 5-ary, 10-ary, and 100-ary signaling.

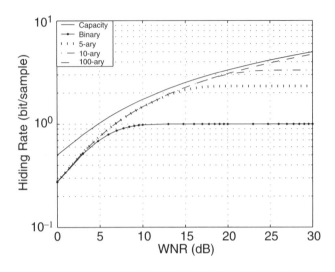

Figure 5-13 Data hiding rates for DM followed by Gaussian-mapping type of postprocessing with binary, 5-ary, 10-ary, and 100-ary signaling.

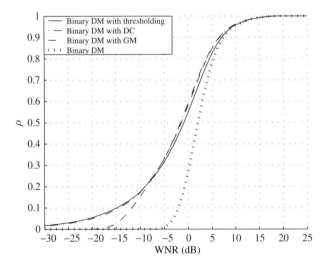

Figure 5-14 The normalized correlation between $\mathbf{W}$ and $\widehat{\mathbf{W}}$ for the considered hiding methods when $P = 10$.

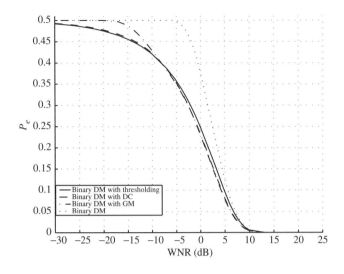

Figure 5-15 The probability of error in detecting W for the considered hiding methods when $P = 10$.

Figs. 5-14 and 5-15. The corresponding embedding/detection parameters for the hiding methods are selected as described in Section 5.2, Eqs. (5.33) and (5.37). The correlation between an embedded binary watermark signal $\mathbf{W}$ and an extracted watermark signal $\widehat{\mathbf{W}}$ is calculated by using Eq. (5.31), and the probability of error in detecting an embedded binary watermark sample is computed by using Eq. (5.36).

The relative performances of the three types of postprocessing obtained for the three criteria, Figs. 5-9, 5-14, and 5-15, are in accord with each other. Thus, the thresholding type of postprocessing performs better when WNR is below approximately −9 dB, and at higher WNRs distortion compensation has better performance. Above −7 dB, Gaussian mapping and distortion compensation have comparable performances, and DM performs well only at the higher WNR range, as expected. Figures 5-16 and 5-17 display the actual simulation results obtained by embedding and detecting binary watermark signals. In Fig. 5-16, the normalized correlation ρ between the embedded vector $\mathbf{W}$ and its extracted version $\widehat{\mathbf{W}}$ is measured, and in Fig. 5-17, the error probability in detecting an embedded

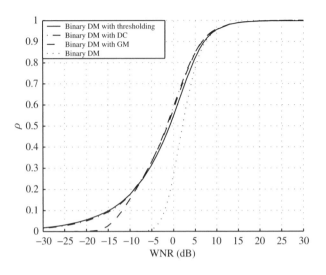

Figure 5-16 The actual measured normalized correlation between embedded $\mathbf{W}$ and extracted $\widehat{\mathbf{W}}$ for the considered hiding methods when $P = 10$.

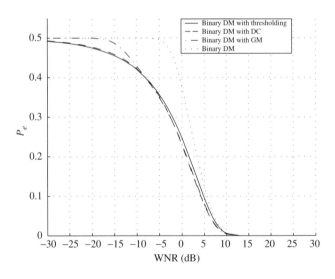

Figure 5-17 The actual measured error probability in detecting W for the considered hiding methods when $P = 10$.

watermark sample W is measured. Both simulation results are in accord with theoretical values computed in Figs. 5-14 and 5-15.

One intuitive way to evaluate the performance characteristics of type I, type II, and type III methods at varying noise levels is by considering the size of decision cells at the detector, as discussed in Section 3.4. For type II methods in the absence of noise, the extracted watermark signals correspond to reconstruction points of the embedding quantizers. Thus, decision cells can collapse to points and the data hider can afford to use higher level signaling without any performance penalty. However, with the increasing noise level, the successful extraction of the embedded watermark signal requires decision cells to be enlarged accordingly. In type III methods, Δ is increased in accordance with the channel noise level σ_Z^2, and the corresponding increase in embedding distortion due to increased Δ is compensated by the postprocessing. Hence, the data hider has the freedom to change the size of the decision cell depending on the noise level. Ultimately, when the noise level is very high, the optimal strategy becomes making the decision regions arbitrarily large, as in type I methods, where even for very high noise levels the detector is able to extract some of the embedded watermark signals.

Advanced Implementations

In all digital communication systems, a general objective is the efficient use of the available resources, that is, bandwidth, power, and complexity, to achieve a specified performance goal. The design of a communication system very often requires tradeoffs among these resources depending on the channel description which quantifies the power limitations, available bandwidth, and nature of the noise and its statistics. In many applications, one of the two primary communications resources, power or bandwidth, is more scarce than the other. This limitation on the communication system is fundamental to the choice of a modulation scheme.

The notion of channel in a communications scenario, defined as the propagating medium between the transmitter and receiver, can be reinterpreted as the host signal in the context of data hiding, as it is the message bearing medium between the embedder and detector. Correspondingly, the power constraints (or SNR) and the available transmission bandwidth of a communications channel can be associated with the amount of embedding distortion (or WNR) and the host signal size (embedding signal size) in the data hiding channel model, respectively. Practical data hiding methods are evaluated based on the performance they deliver over varying WNR levels at the given signal size N and the degree of complexity involved in their implementations. Due to the nature of the applications, however, data hiding methods are required to trade off complexity for performance, depending on the signal size N.

For the case where the host signal size is large, *spread transforming* can be employed. Spread transforming (ST) technique is inspired by spread-spectrum (SS) modulation scheme [19]. Spread-spectrum systems are most generally viewed as power limited systems (bandwidth efficiency is not of primary concern) since the bandwidth occupancy of transmitted information signal is much wider than and independent of the bandwidth that is intrinsically needed. In SS communications, the information signal is spread in bandwidth prior to transmission and then despread in bandwidth by the same factor at the receiver. While this scheme keeps the total transmitted signal power unchanged, it reduces the power spectral density of the narrow-band noise signal introduced during transmission through despreading. Dually, ST increases the embedding distortion to noise distortion ratio (WNR) at the extractor by sacrificing the signal size N. The concept of optimal spreading factor in ST is addressed by Ramkumar *et al.* in [21] and by Eggers *et al.* in [62] independently.

On the contrary, when the signal size is small, *multiple codebook data hiding* method can be used [63], [64]. In strictly bandwidth limited communication systems, an efficient error control scheme, based on set-partitioning of an expanded constellation, is employed to improve the performance (i.e., trellis coded modulation). A key to the success of this scheme is the increased minimum distance between the codewords which enables more reliable decoding of the sent message. Similar to set-partitioning principle of trellis encoding, quantization based methods employ a periodic constellation where each information symbol is associated with a higher minimum distanced subconstellation. In addition, the multiple codebook data hiding method generates a set of codewords for each message to be hidden, thereby expanding the constellation by the number of codebooks, and picks the codeword that adapts to host signal best. In multiple codebook data hiding, the detection performance is improved due to the ability to embed the message signal at a lower embedding distortion.

6.1 Spread Transforming

The underlying idea of spread transforming is to embed the watermark signal into a projection of the host signal and generate the stego signal

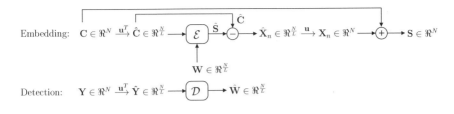

Figure 6-1 Embedding and detection with spread transforming.

by spreading the corresponding lower-dimensional embedding distortion over the high-dimensional host signal. In spread transforming, a pseudo-random vector $\mathbf{u}$ of size L with unit norm, $1 = \mathbf{u}^T\mathbf{u}$, is designed as the spreading vector and made known to both embedder and detector. The embedding and detection operations are performed as follows.

At the embedder, the host signal vector $\mathbf{C} \in \Re^N$ is split into sub-vectors of length L such that $\mathbf{C}^T = [\mathbf{C}_1^T, \ldots, \mathbf{C}_{\frac{N}{L}}^T]$, where $\mathbf{C}_i \in \Re^L$. Each block of data is projected onto $\mathbf{u}$ to generate the projected host signal $\widehat{\mathbf{C}}^T = [\widehat{C}_1, \ldots, \widehat{C}_{\frac{N}{L}}]$, where $\widehat{\mathbf{C}} \in \Re^{\frac{N}{L}}$ and

$$\widehat{C}_i = \mathbf{C}_i^T\mathbf{u}, \quad i = 1, \ldots, \frac{N}{L}. \tag{6.1}$$

Then, the watermark signal $\mathbf{W} \in \Re^{\frac{N}{L}}$, corresponding to a message index, is embedded into $\widehat{\mathbf{C}}$, $\widehat{\mathbf{S}} = \mathcal{E}(\widehat{\mathbf{C}}, \mathbf{W})$. The stego signal $\mathbf{S}^T = [\mathbf{S}_1^T, \ldots, \mathbf{S}_{\frac{N}{L}}^T]$ is generated from $\widehat{\mathbf{S}}^T = [\widehat{S}_1, \ldots, \widehat{S}_{\frac{N}{L}}]$ as

$$\mathbf{S}_i = \mathbf{C}_i + (\widehat{S}_i - \widehat{C}_i)\mathbf{u}, \quad i = 1, \ldots, \frac{N}{L}. \tag{6.2}$$

Similarly, at the detector, the received signal is partitioned into blocks, $\mathbf{Y}^T = [\mathbf{Y}_1^T, \ldots, \mathbf{Y}_{\frac{N}{L}}^T]$, and each block of data is projected onto $\mathbf{u}$, $\widehat{\mathbf{Y}} = [\widehat{Y}_1, \ldots, \widehat{Y}_{\frac{N}{L}}]$, where $\widehat{Y}_i = \mathbf{Y}_i^T\mathbf{u}$. This is followed by the detection of the hidden signal, $\mathcal{D}(\widehat{\mathbf{Y}})$. Figure 6-1 depicts the embedding and detection operations with spread transforming.

With spreading, the bandwidth is reduced by a factor of L, from N to $\frac{N}{L}$, as $\frac{N}{L}$ coefficients are information embedded. However, the embedding distortion is spread over all of the N coefficients. On the other hand, the distortion introduced into the host signal is $\frac{N}{L}P = \|\widehat{\mathbf{S}} - \widehat{\mathbf{C}}\|^2$, which would be NP without the spreading. Therefore, the hider can afford to increase the embedding power by a factor of L. At the embedder, this is reflected in an increase in the distance between the reconstruction points of the embedding quantizers (when scalar quantization is considered, spreading with a factor of L leads to an increase in Δ by a factor of $\sqrt{L}$, i.e., $LP = (\sqrt{L}\Delta)^2/12$, where $P = \Delta^2/12$ is the embedding distortion per coefficient). Therefore, the system operates at a higher WNR level. An alternate interpretation of the gain due to spreading is that the stego signal can be distorted only by the component of the noise that is in the direction of the vector $\mathbf{u}$, which improves the robustness against noise.

The spread transforming method can be generalized to include noninteger spreading factors by adopting a transform domain embedding/detection approach in which each basis vector of the transform basis is treated as a spreading vector. Let $\mathbb{U} \in \mathfrak{R}^{L \times L}$ be a unitary transformation matrix, $\mathbb{I} = \mathbb{U}^T\mathbb{U}$, where $\mathbb{I}$ is an $L \times L$ identity matrix, and the host signal vector $\mathbf{C} \in \mathfrak{R}^N$ be mapped to the matrix $\mathbb{C} \in \mathfrak{R}^{L \times \frac{N}{L}}$, by arranging its coefficients into L rows and $\frac{N}{L}$ columns, $\mathbb{C} = [\mathbf{C}_1; \ldots ; \mathbf{C}_L]$, where $\mathbf{C}_i \in \mathfrak{R}^{1 \times \frac{N}{L}}$. Let $\widehat{\mathbb{C}}$ represent the unitary transformation of $\mathbb{C}$ as

$$\widehat{\mathbb{C}} = \mathbb{U}\mathbb{C} \tag{6.3}$$

where $\widehat{\mathbb{C}} = [\widehat{\mathbf{C}}_1; \ldots ; \widehat{\mathbf{C}}_L]$ and $\widehat{\mathbf{C}}_i \in \mathfrak{R}^{1 \times \frac{N}{L}}$. In other words, the coefficients of the host signal vector are broken down into L channels, each consisting of $\frac{N}{L}$ coefficients. The watermark signal $\mathbf{W} \in \mathfrak{R}^{\frac{N}{L}}$ is embedded in the coefficients of designated channel(s), i.e., $\widehat{\mathbf{C}}_1, \ldots, \widehat{\mathbf{C}}_L$.

For the general case, let's assume that $\mathbf{W}$ is embedded into ith channel coefficients. This yields the embedded signal $\hat{\mathbf{S}}_i = \mathcal{E}(\widehat{\mathbf{C}}_i, \mathbf{W})$ at the ith channel, while the transform coefficients in the rest of the channels are not changed. Then the transformed and embedded signal $\hat{\mathbb{S}} = [\widehat{\mathbf{C}}_1; \ldots ; \hat{\mathbf{S}}_i; \ldots ; \widehat{\mathbf{C}}_L]$ is inverse transformed as

$$\mathbb{S} = \mathbb{U}^T \hat{\mathbb{S}} \tag{6.4}$$

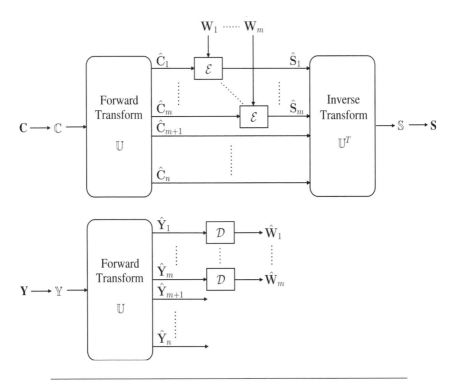

Figure 6-2 Embedding and detection of $\mathbf{W}^T = [\mathbf{W}_1^T, \ldots, \mathbf{W}_m^T]$ into $\mathbf{C}$ with the spreading gain $L = \frac{n}{m}$.

and mapped to the stego signal vector $\mathbf{S}$. At the detector, the embedded signal is extracted from the stego channel(s) obtained by segmenting and transforming the received signal $\mathbf{Y}$. Although only particular transform coefficients are used for data hiding, the resulting embedding distortion, in the transform domain, is spread over all samples in the signal domain. This enables the hider to exploit the bandwidth vs WNR trade-off at the detector by selecting the spreading factor by choosing $\mathbb{U}$. The spreading factor, in this case, is the ratio of the total number of channels to the number of channels used for data hiding. In order to obtain a spreading factor of $L = \frac{n}{m}$, where L may also be a rational number, m channels of an $n \times n$ unitary transform of the host signal ($\mathbb{C} \in \Re^{n \times \frac{N}{n}}$) are information embedded. Figure 6-2 illustrates this scenario, in which the first m channels of $\widehat{\mathbb{C}}$ are used for hiding the watermark signal $\mathbf{W}^T = [\mathbf{W}_1^T, \ldots, \mathbf{W}_m^T]$, where $\mathbf{W}_i \in \Re^{\frac{N}{n}}$.

The effect of spread transforming on the data hiding rate of a method can be determined in terms of N and WNR. The capacity of any communication scheme, in general, can be expressed in terms of its bandwidth and SNR. Therefore, the data hiding capacity can be formulated as $C = Nf(\text{WNR})$. Due to the trade-off between N and WNR, the capacity with spread transforming takes the form of $C_S = \frac{N}{L}f(L \times \text{WNR})$. Thus, the optimal spreading factor L for a given method can be found from the measured C through maximizing C_S. It should be noted that if the capacity formulation of a scheme is such that the linear increase in the WNR can compensate for the linear reduction in N, then spread transforming offers an improvement in performance. As the performance drop in type II and type III methods are exponential in WNR, spreading becomes an efficient tool by enabling them to operate at higher WNR levels, where they perform reasonably well. However, for type I schemes and the upper bound (optimal type III scheme), where all variables are assumed to be Gaussian, the fall in the hiding rate is logarithmic, $\frac{1}{2} \log_2(\frac{\text{WNR}}{\text{WNR} \times \text{DWR} + 1})$ and $\frac{1}{2} \log_2(1 + \text{WNR})$, respectively. Consequently, the optimal spreading factor L that maximizes $\frac{1}{2L} \log_2(\frac{L \times \text{WNR}}{L \times \text{WNR} \times \text{DWR} + 1})$ or $\frac{1}{2L} \log_2(1 + L \times \text{WNR})$ is computed as 1.

The hiding rate vs robustness curves of DM and type III methods with spread transforming, computed using the results of Fig. 5-9, are displayed in Fig. 6-3. When compared with the hiding capacity, the hiding rates corresponding to DM and the type III implementations of DM with the Gaussian-mapping type of postprocessing improved remarkably in the low WNR range. With spread transforming, distortion-compensation and Gaussian-mapping types of processing deliver slightly better performances than the thresholding type. This is not surprising because the improvement with spreading depends on the performance of the scheme at higher WNRs where distortion-compensation and Gaussian-mapping types of postprocessing were seen to perform better than thresholding type of postproccessing (see Section 5.3). Measured spreading factors for the methods are shown in Fig. 6-4. However, one should be careful because very large spreading factors enable large embedding distortions, i.e., increased Δ values, and this may violate the assumption that the host signal is uniformly

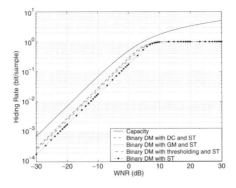

Figure 6-3   Improvement in the hiding rate of type II and type III methods when $P = 10$.

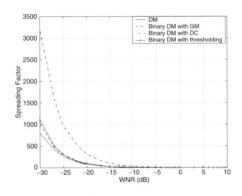

Figure 6-4   Corresponding spreading factors.

distributed over all quantization cells. Therefore, large spreading factors may not be practically feasible as the embedding operation becomes dependent on the statistics of the host signal.

6.2   Multiple Codebook Data Hiding

When the embedding signal size N is small, multiple codebook data hiding can be used to embed the watermark signal at lower embedding distortion levels. The distortion P introduced into host signal $\mathbf{C}$ due to the embedding operation is computed over all stego signal coefficients as $P = \frac{1}{N}\|\mathbf{X}_n\|^2$. Assuming that the pdf of the host signal is smooth enough, such that it can be considered uniformly distributed over all quantization intervals, the distortion introduced into each host signal sample C has the statistics of X_n, Eqs. (5.16), (5.22), and (5.25). In other words, the distortion P is a random variable and its distribution approximates $\mathcal{N}(\sigma_{X_n}^2, \frac{\sigma_P^2}{N})$, where

$$\frac{\sigma_P^2}{N} = \frac{1}{N} \int_{-\infty}^{\infty} x_n^4 f_{X_n}(x_n)\,dx_n - (\sigma_{X_n}^2)^2. \tag{6.5}$$

Accordingly, when N is large, the distortion P introduced into the host signal becomes $\sigma_{X_n}^2$. However, when N is small, P varies around the mean $\sigma_{X_n}^2$ depending on the distribution of X_n and the signal size N. The variation in the embedding distortion becomes more significant with the decreasing value of N. Therefore, embedding in a host signal with limited signal size requires a more careful selection of embedding and detection parameters. In general, embedding/detection parameters are optimized to maximize the performance at the given noise level σ_Z^2 and the permitted distortion $\sigma_{X_n}^2$ as described in Section 5.2.3. Therefore, implicitly, a very large embedding signal size N is assumed. Embedding and detection with the parameters obtained through an optimization procedure that disregards this aspect of the problem may cause the data hiding method to operate on a lower hiding rate vs robustness curve due to the variation in the embedding distortion with respect to N.

Figures 6-5 and 6-6 display the hiding rates corresponding to binary DM with thresholding and distortion-compensation types of postprocessing for various N values when the embedding distortion deviates from the mean $\sigma_{X_n}^2$ by five times the standard deviation, $P = \sigma_{X_n}^2 - 5\sqrt{\sigma_P^2/N}$. As displayed in the figures, with decreasing N, the hiding rate drops in both cases. However, since X_n corresponding to the distortion-compensation type of postprocessing has higher variance around the mean, the reduction in rate is more drastic. These results indicate that given two host signals with similar statistics, if the same watermark signal is embedded in both signals using the same parameters, the resulting distortion due to

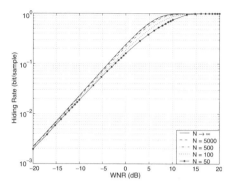

Figure 6-5 Hiding rates corresponding to binary DM with thresholding for various N when $P = \sigma_{X_n}^2 - 5\sqrt{\sigma_P^2/N}$.

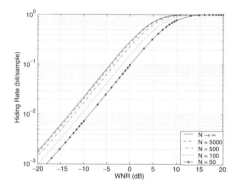

Figure 6-6 Hiding rates corresponding to binary DM with distortion compensation for various N when $P = \sigma_{X_n}^2 - 5\sqrt{\sigma_P^2/N}$.

embedding may differ significantly for the two signals depending on size N. Therefore, more sophisticated optimization techniques are needed for determining the embedding/detection parameters for limited N. An obvious approach is to fine-tune the parameters obtained with the assumption of large N, such that the resulting distortion is neither above nor below the permitted distortion level. The question now is, can we do better? Can the fact that the embedding distortion has a large variance be utilized to improve the performance of data hiding? We shall soon see that this is possible. The multiple codebook data hiding method exploits this phenomenon by choosing a transformation of **C** that yields the minimum embedding distortion when **W** is embedded. The ability to embed a watermark signal at a lower embedding distortion, rather than at the permitted distortion level, is translated into more robust embedding of the watermark signal.

The essence of the method is depicted in Figs. 6-7 and 6-8, where the embedding signal size is 2. In both cases, one of the binary symbols is embedded into a signal vector **c** composed of two signal samples, $\mathbf{c} = (c_1, c_2)$, using either a two-dimensional lattice or two unidimensional lattices. The lattice points or the reconstruction points associated with each binary sample is marked by $\times$ and $\circ$ symbols. The embedding operation is the translation of the vector **c** to the nearest centroid associated with the symbol to be embedded. The decision regions in Figs. 6-7 and 6-8 determine the sustainable amount of noise that does not impair the detection performance.

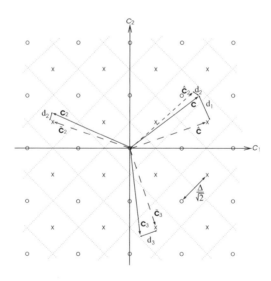

Figure 6-7   Depiction of embedding a binary symbol into the host signal $\mathbf{c} = (c_1, c_2)$ and into its two transformations using a 2-D lattice.

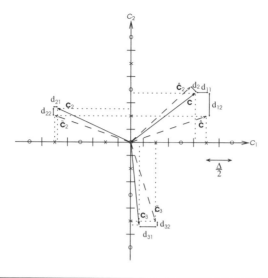

Figure 6-8 Depiction of embedding two binary symbols into the host signal vector $\mathbf{c} = (c_1, c_2)$ and into its two transformations using uniform scalar quantizers.

In the considered cases, the binary symbol corresponding to $\times$ is embedded into $\mathbf{c}$ and into two of its transformed (rotated) versions $\mathbf{c}_2$ and $\mathbf{c}_3$. The embedding distortions between the signal pairs $(\mathbf{c}, \hat{\mathbf{c}})$, $(\mathbf{c}_2, \tilde{\mathbf{c}}_2)$, and $(\mathbf{c}_3, \tilde{\mathbf{c}}_3)$ are measured, in terms of Euclidean distance, as d_1, d_2, and d_3, respectively, as displayed in Fig. 6-7. Similarly, in Fig. 6-8 the resulting embedding distortions are measured as $\sqrt{d_{11}^2 + d_{12}^2}$, $\sqrt{d_{21}^2 + d_{22}^2}$, and $\sqrt{d_{31}^2 + d_{32}^2}$. When $\tilde{\mathbf{c}}_2$ and $\tilde{\mathbf{c}}_3$ are inverse transformed, one can observe that the distortions introduced into $\mathbf{c}$ due to three embedding operations are not the same. For both of the cases depicted in Figs. 6-7 and 6-8, $\hat{\mathbf{c}}_2$ (inverse transformed $\tilde{\mathbf{c}}_2$) yields the smallest embedding distortion, d_2. It is important to note that the amount of embedding distortion, due to embedding into transformations of $\mathbf{c}$, $\mathbf{c}_2$, and $\mathbf{c}_3$, remains the same in magnitude after the inverse transformation because the transformation is assumed to be unitary or energy preserving. One can now easily see that with the added computational complexity of transformations, a binary symbol can be embedded into $\mathbf{c}$ at a smaller embedding distortion level. Multiple codebook hiding incorporates these savings in embedding distortion with type III hiding methodology.

Type III methods, as described earlier, are derived from type II methods by increasing the distance between the reconstruction points and introducing a processing distortion that is also a function of the expected noise level. In type III methods, the resulting increase in the embedding distortion, due to the increased separation of the reconstruction points, is reduced to the permitted amount by postprocessing, while performance is maximized at the expected noise level [49]. In other words, the distortion introduced into $\mathbf{C}$ due to the embedding operation is limited to the permitted amount P by proper selection of the separation between the reconstruction points (Δ) and the amount of processing distortion ($\sigma_{\tilde{X}_t}^2$). The Δ and $\sigma_{\tilde{X}_t}^2$ values that yield the distortion P are not unique, and in order to maintain a fixed distortion level of P, an increase or decrease in either of Δ or $\sigma_{\tilde{X}_t}^2$ values should be followed by the other in the same manner. *Since the employment of transformations enables embedding at lower distortion levels, the difference between the permitted and actual embedding distortions can be utilized by the type III embedder to either reduce the $\sigma_{\tilde{X}_t}^2$ value at the given Δ or further increase the Δ value at the fixed $\sigma_{\tilde{X}_t}^2$.* Both actions lead to an improvement in the detection performance.

Employing multiple codebooks resembles the optimal binning technique in the manner that the size of each bin is increased from 1 to the number of codebooks. Therefore, for a message to be transmitted, the embedder generates a set of codewords and chooses the best among them. Correspondingly, the detector has to search over all codebooks for a successful extraction of the message. Modifying the multiple codebook hiding method by assigning one of the codebooks for embedding and detection while discarding the others reduces it to a type III method. Due to this freedom in selecting one of the many codebooks being utilized, the method is referred to as multiple codebook hiding.

In multiple codebook hiding, each codeword is generated from a unitary transformation of the host signal. From this point of view, the design of the ideal codebook requires the derivation of the optimal transform basis for embedding and detection (at both the embedder and the detector). This is an impractical task considering the dependency on the host signal. (In Figs. 6-7 and 6-8, where $N = 2$, this refers to the transformation that translates $\mathbf{c}$ to a point that coincides with one of the $\times$ points.) Therefore, rather than computing the optimal transformation basis, a set of transformation bases is selected with the intention that for a given host signal, some of the bases will yield codewords similar to those of the optimal transformation. Thus, the use of multiple codebooks provides the embedder with a freedom in choosing the best among a number of suboptimal codewords. However, when $N \longrightarrow \infty$, for any $\mathbf{C}$, the embedding distortion converges to the expected value, $P \longrightarrow \sigma_{X_n}^2$, and the multiple codebook hiding method does not provide any advantage over single codebook hiding. (In other words, with the increasing N, all transformations of $\mathbf{C}$ become equally preferable for embedding, as they all yield the same distortion.) On the other hand, the detector should be able to differentiate the correct transformation from among all transformations of the received signal in order to successfully detect the embedded message. Apparently, such a detection of the message is more prone to errors. Ultimately, the question to be answered is whether at a fixed N and permitted embedding distortion, the improvement in the detection performance due to the ability to increase the Δ (or to reduce the $\sigma_{X_l}^2$) can compensate for the additional detection errors due to the uncertainty in the transform basis used for embedding. It is shown that for the AWGN channel, Gaussian distributed host signal, and squared error distortion measure, the increase in probability of error due to use of multiple codebooks is compensated by a reduction (in probability of error) due to the embedder's ability to adapt the codeword to the host signal.

6.2.1 A Channel Model for Multiple Codebook Data Hiding

In the multiple codebook data hiding scenario, the information hider and extractor share two sets of information. One is the set of sequences $\mathbf{W}_1, \ldots, \mathbf{W}_M \in \mathfrak{R}^N$ that are associated with M distinct messages. The other is the set of L, $N \times N$, unitary transform bases, i.e.,

$$\mathbb{I} = \mathbb{T}_i^T \mathbb{T}_i, \quad i = 1, \ldots, L \tag{6.6}$$

where $\mathbb{I}$ is the $N \times N$ identity matrix and ^{T} denotes the matrix transpose operation. The overall data hiding system is outlined in Eqs. (6.7) through

(6.12) in an additive model as follows:

$$\mathcal{W} : m \longrightarrow \mathbf{W}_m, \tag{6.7}$$

$$\widehat{\mathbf{S}}_k = \mathcal{E}(\mathbb{T}_k \mathbf{C}, \mathbf{W}_m), \quad 1 \leq k \leq L, \tag{6.8}$$

$$\mathbf{S}_k = \mathbb{T}_k^T \widehat{\mathbf{S}}_k, \tag{6.9}$$

$$\mathbf{Y} = \mathbf{S}_k + \mathbf{Z} = \mathbf{C} + \mathbf{X}_{n_k} + \mathbf{Z}, \tag{6.10}$$

$$\widehat{\mathbf{W}}_m^i = \mathcal{D}(\mathbb{T}_i \mathbf{Y}), \quad i = 1, \ldots, L, \tag{6.11}$$

$$\mathcal{W}^{-1} : \widehat{\mathbf{W}}_m^i \longrightarrow \hat{m}. \tag{6.12}$$

In the model, $\mathbf{C}$ is the *iid* Gaussian distributed host signal with the marginal $C \sim \mathcal{N}(0, \sigma_C^2)$, $\mathbf{X}_n = \mathbf{X}_{n_k}$ is the distortion introduced by the type III embedder (type III codeword, Section 5.2), and $\mathbf{Z}$ is the AWGN vector, where $Z \sim \mathcal{N}(0, \sigma_Z^2)$. One selection criterion for $\mathbb{T}_i$, $i = 1, \ldots, L$, is to require that the transformations of a random signal vector $\mathbf{r}$ be maximally separated from each other in $\mathfrak{R}^N$ with respect to a predesignated distance measure. For squared error distortion measure, selection of $\mathbb{T}_1, \ldots, \mathbb{T}_L$ is based on the maximization of the following criterion

$$E[\|\mathbb{T}_k \mathbf{r} - \mathbb{T}_i \mathbf{r}\|^2], \quad 1 \leq i, k \leq L \text{ and } i \neq k \tag{6.13}$$

where the expectation is performed over all $\mathbf{r} \in \mathfrak{R}^N$. Among the L unitary transformations $\mathbf{C}_i = \mathbb{T}_i \mathbf{C}$, $i = 1, \ldots, L$, the embedder picks the one that is expected to yield the highest detection statistics at the permitted embedding distortion. Assuming that k is the index of the selected transform basis, the sequence $\mathbf{W}_m$, corresponding to the message indexed by m, $1 \leq m \leq M$, is embedded in the $\mathbb{T}_k$ transformation of the host signal, $\mathbf{C}_k$. Then, the stego signal in the transform domain $\widehat{\mathbf{S}}_k$ is inverse transformed to the signal domain $\mathbf{S}_k$. Uninformed of the particular transform $\mathbb{T}_k$ used for embedding, the detector generates L transformations of the received signal $\mathbf{Y}$ and detects the hidden message $\hat{m}$ in a blind manner. With the use of multiple codebooks, the choice of T_k determines the codeword $\mathbf{X}_{n_k}$ among codewords $\{\mathbf{X}_{n_1}, \ldots, \mathbf{X}_{n_L}\}$. Therefore, the embedding operation can be viewed as a vectorial operation in which the embedder chooses one of the

TABLE 6-1
Notations Used in This Chapter

$\mathbf{C}$	Host signal vector
$\mathbf{X}$	Codeword
$\mathbf{S}$	Information hidden signal vector
$\mathbf{Z}$	Channel noise vector
$\mathbf{Y}$	Distorted $\mathbf{S}$
$\mathbf{W}_m$	Watermark signal vector corresponding to message m to be conveyed
$\widehat{\mathbf{W}}_m$	Extracted signal vector when $\mathbf{W}_m$ is embedded
$\rho_{m,j}$	The normalized correlation between $\widehat{\mathbf{W}}_m$ and $\mathbf{W}_j$
$d_{m,j}$	The mean squared distance between $\widehat{\mathbf{W}}_m$ and $\mathbf{W}_j$
$\widehat{\mathbf{W}}_m^i$	Extracted signal vector from $\mathbb{T}_i$ transformation of $\mathbf{Y}$ when $\mathbf{W}_m$ is embedded
$\widetilde{\mathbf{W}}_m^i$	Extracted signal vector from $\mathbb{T}_i$ transformation of $\mathbf{S}$ when $\mathbf{W}_m$ is embedded
$\rho_{m,j}^i$	The normalized correlation between $\widehat{\mathbf{W}}_m^i$ and $\mathbf{W}_j$
$\tilde{\rho}_{m,j}^i$	The normalized correlation between $\widetilde{\mathbf{W}}_m^i$ and $\mathbf{W}_j$
$d_{m,j}^i$	The mean squared distance between $\widehat{\mathbf{W}}_m^i$ and $\mathbf{W}_j$
$\tilde{d}_{m,j}^i$	The mean squared distance between $\widetilde{\mathbf{W}}_m^i$ and $\mathbf{W}_j$

L codewords based on the given host signal $\mathbf{C}$ and the message m to be conveyed.

Figure 6-9 displays codeword generation for multiple codebook hiding. Compared with Fig. 3-7, the main difference is that for a message index m, L number of codewords are generated by embedding $\mathbf{W}_m$ into $\mathbb{T}_1, \ldots, \mathbb{T}_L$ transformations of $\mathbf{C}$. Consequently, the embedder chooses the best one among the codewords $\mathbf{X}_{1,m}, \ldots, \mathbf{X}_{L,m}$.

Table 6.1 lists all the notations used in this analysis in addition to the previous notation; i.e., the vectors are denoted by boldfaced characters, the random variables and their realizations are symbolized by the capital letters and the corresponding lowercase letters, respectively. For the general case, all signals are assumed to be random vectors of size N. However, in some of the derivations, individual random variables are used for the sake of simplicity. In such cases, vector extensions are straightforward due to *iid* assumption.

The most crucial step in multiple codebook data hiding is the selection of the transformation basis $\mathbb{T}_k$, $1 \leq k \leq L$, which yields the

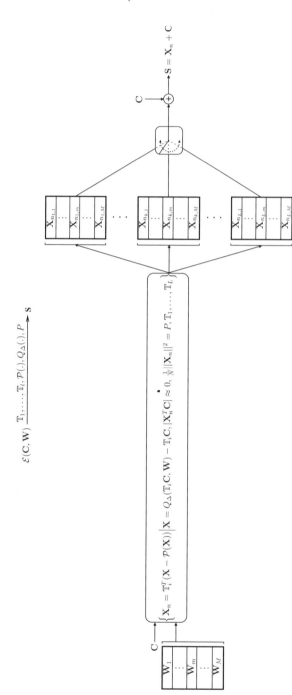

Figure 6-9 Encoding of message index m using multiple codebooks.

codeword that adapts to the host signal $\mathbf{C}$ best at the permitted embedding distortion. For this, the watermark signal $\mathbf{W}_m$ is embedded into L transformations of the host signal, $\mathbf{C}_i = \mathbb{T}_i \mathbf{C}$, $i = 1, \ldots, L$, consecutively. Note that in a type III method, embedding and detection functions are not inverses of each other; the signal $\mathbf{W}_m$ embedded into $\mathbf{C}_i$ will differ from the corresponding extraction $\widetilde{\mathbf{W}}_m^i$ due to the processing distortion $\mathbf{X}_t$, $\mathcal{D}(\mathcal{E}(\mathbf{C}, \mathbf{W}_m)) \neq \mathbf{W}_m$. Therefore, the embedder can decide on the transformation basis by measuring the similarity (or dissimilarity) between $\mathbf{W}_m$ embedded in all transformations of $\mathbf{C}$ and the corresponding extractions $\widetilde{\mathbf{W}}_m^i$ through computing and comparing normalized correlations, $\tilde{\rho}_{m,m}^i$, or mean squared distances, $\tilde{d}_{m,m}^i$. If the decision on the transform basis is made using correlation, the *maximum correlation* criterion, the value of index i that yields the highest correlation $\tilde{\rho}_{m,m}^i$, is chosen as the index of the best transformation basis $\mathbb{T}_k$, $k = \arg\max_i \left(\tilde{\rho}_{m,m}^i \right)$ for $\tilde{\rho}_{m,m}^i = (\mathbf{W}_m^T \widetilde{\mathbf{W}}_m^i)/(\|\mathbf{W}_m\| \, \|\widetilde{\mathbf{W}}_m^i\|)$. Alternately, if squared error distance is used as the decision metric, the *minimum distance* criterion, the embedder picks the transform basis $\mathbb{T}_k$ that yields the smallest mean squared distance between $\mathbf{W}_m$ and $\widetilde{\mathbf{W}}_m^i$, $k = \arg\min_i\{\tilde{d}_i\}$, $i = 1, \ldots, L$, where $\tilde{d}_i = \frac{1}{N} \|\mathbf{W}_m - \widetilde{\mathbf{W}}_m^i\|^2$.

Such a selection of the transformation basis can be justified as follows. In order to embed a signal in a host signal, the embedder has to determine the optimal embedding parameters depending on the postprocessing employed (i.e., (Δ, β) for thresholding, (Δ, α) for distortion compensation). These parameters are computed in advance for the permitted embedding distortion (P_E) and the given channel noise (σ_Z^2) levels assuming that N is very large and host signal is uniformly distributed in each quantization interval. It should be noted that the embedding parameters computed using the optimization criteria described in Section 5.2 are valid when N is relatively large. However, due to limitation on the size N, the embedding distortion P introduced into $\mathbf{C}$ by using the optimal embedding parameter values differs from P_E. Therefore, the embedder has to fine-tune those parameters for the given host signal in order to comply with P_E. Since Δ is also revealed to the extractor, it should remain the same for all embedding operations while processing distortion due to the choice of β or α may vary for each embedding. As discussed earlier, β and α designate the amount of processing distortion applied on the type II codeword due to the postprocessing. Ultimately, when $\beta = \Delta$ or $\alpha = 1$ no postprocessing is performed, and therefore, embedded and extracted watermark signals are the same. On the other hand, when embedding of $\mathbf{W}_m$ with $\beta < \Delta$ and $\alpha < 1$ is considered, the extracted signal $\tilde{\mathbf{W}}_m$ will be distorted at various levels depending on the amount of processing distortion. Thus, the correlation (respectively the distance) between the embedded and extracted signals reduces (increases) with decreasing β or α.

The sent message is detected from the received signal $\mathbf{Y}$ without knowing which of the L transformation bases is used for embedding. Hence, the extractor tries all preset transformations of $\mathbf{Y}$ and extracts signals $\widehat{\mathbf{W}}_m^i = \mathcal{D}(\mathbb{T}_i \mathbf{Y})$. Then, the set of extracted signals $\{\widehat{\mathbf{W}}_m^1, \ldots, \widehat{\mathbf{W}}_m^L\}$, of which only $\widehat{\mathbf{W}}_m^k$ is a valid extraction, is compared with the set of watermark signals $\{\mathbf{W}_1, \ldots, \mathbf{W}_M\}$ by computing the normalized correlations, $\rho_{m,j}^i$, or mean squared distances, $d_{m,j}^i$, where $i = 1, \ldots, L$ and $j = 1 \ldots, M$, depending on the decision metric used at the embedder. Among all (i, j) index pairs, the j index of the pair that maximizes $\rho_{m,j}^i$ or minimizes $d_{m,j}^i$ is the index of the detected message $\hat{m}$, $\hat{m} = \arg_j \max_{i,j} \left(\rho_{m,j}^i \right)$ or $\hat{m} = \arg_j \min_{i,j} \left(d_{m,j}^i \right)$.

Figure 6-10 displays an L codebook embedding and detection scheme. In the block diagram, $\mathbf{W}$ is the watermark signal corresponding to message index m. The decision block b_E, at the embedder, decides on the best transform basis $\mathbb{T}_i$, $1 \le i \le L$, to be employed for embedding using one of the decision metrics. Then, it transmits the stego signal corresponding

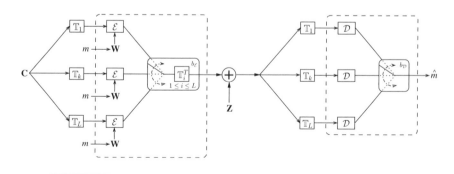

Figure 6-10   Multiple codebook embedding and detection.

to $\mathbf{W}$ and $\mathbf{C}$. At the detector, b_D detects the message with index $\hat{m}$ by computing the correlations or distances between the extracted signals and the set of watermark signals. A detection error occurs whenever m and $\hat{m}$ are not the same.

In multiple codebook hiding, as mentioned earlier, the embedder is able to better adapt the codeword to the host signal. However, this improvement at the embedder is accompanied by an increase in the probability of detection error. This error is due to the two sources of noise: the channel noise and the interference from the other transformations. When extraction is made from the correct transformation of the received signal, the sent message may still be falsely detected due to the channel's distortion affecting the stego signal. This scenario is the same as the detection error in single codebook hiding. However, for the multiple codebook case, the error may also be due to the interference from the other $L - 1$ transformations. This occurs when detection of a message is obtained from a transformation of the received signal *other* than the transformation used at the embedder. This error is independent of the channel noise and can be minimized by the proper selection of the transformation bases.

Type III schemes like binary DM with thresholding and distortion-compensation types of postprocessing, employing soft-decision-rule–based detectors, are incorporated with the multiple codebook data hiding technique. In Sections 6.2.2–6.2.5, single and multiple codebook hiding methods utilizing *maximum correlation* and *minimum distance* criteria are studied. Their probability of error performances are also calculated and presented in these sections.

6.2.2   Single Codebook Data Hiding Based on the Maximum Correlation Criterion

Let $\mathbf{W}_m^T = [W_{m_1}, \ldots, W_{m_N}]$ be a length-N *iid*, zero-mean, binary random vector corresponding to message m, and $\widehat{\mathbf{W}}_m^T = [\widehat{W}_{m_1}, \ldots, \widehat{W}_{m_N}]$ be the extracted real-valued signal at the detector. Since the embedding and detection processes are memoryless, and both host signal and channel noise are white, $\widehat{\mathbf{W}}_m$ is an *iid*, zero mean random vector. For the single codebook case, the embedder employs an $M \times N$–sized codebook composed of M length-N codewords. A detection error is due to $\widehat{\mathbf{W}}_m$ having the highest correlation with any of $\{\mathbf{W}_1,\ldots,\mathbf{W}_M\}$ other than $\mathbf{W}_m$. Then, an event E_j that the detector will pick $\hat{m}$ as the detected message instead of m is denoted as

$$E_j = \{p(\rho_{m,j} \geq \rho_{m,m})\}, j = 1,\ldots,M \text{ and } j \neq m. \tag{6.14}$$

The event E that the detector makes a detection error is expressed as

$$E = \bigcup_{j=1, j\neq m}^{M} E_j. \tag{6.15}$$

Hence, the probability of error for single codebook–based data hiding, P_e^{one}, is expressed as

$$P_e^{one} = Pr\{E\} \leq \sum_{j=1, j\neq m}^{M} Pr\{E_j\}. \tag{6.16}$$

Using Eq. (6.14), the upper bound for P_e^{one} can be rearranged as

$$P_e^{one} \leq \sum_{j=1, j\neq m}^{M} p(\rho_{m,j} \geq \rho_{m,m}). \tag{6.17}$$

In Eq. (6.17), $\rho_{m,j}$ and $\rho_{m,m}$ are random variables that are equivalent to random variables ρ_{ind} and ρ_{dep}, respectively, in their statistics. The relationship of $\rho_{m,j}$, $1 \leq m,j \leq M$, with ρ_{ind} and ρ_{dep} is explained in the following subsections. Based on those results, the pdf of random variable $\rho_{m,j}$ can be generalized as

$$\rho_{m,j} \sim \begin{cases} \mathcal{N}(0,\frac{1}{N}), & \text{if } m \neq j \\ \rho_{dep}, & \text{if } m = j. \end{cases} \tag{6.18}$$

Assuming that m is the index of the transmitted message for all the cases, the first subscript, m, of $\rho_{m,j}$ can be dropped for the sake of simplicity. Thus, Eq. (6.17) can be rewritten using Eq. (6.18) as

$$P_e^{one} \leq \sum_{j=1, j \neq m}^{M} \int_{-\infty}^{\infty} \int_{-\infty}^{\infty} f_{\rho_j}(\rho_j \geq \rho_m) f_{\rho_m}(\rho_m) \, d\rho_j \, d\rho_m, \qquad (6.19)$$

$$\leq \sum_{j=1, j \neq m}^{M} \int_{-\infty}^{\infty} \left(\int_{\rho_m}^{\infty} f_{\rho_j}(\rho_j) d\rho_j \right) f_{\rho_m}(\rho_m) \, d\rho_m. \qquad (6.20)$$

The inner integral in Eq. (6.20) can be expressed in terms of Gaussian Q-function, i.e., $Q(x) = \frac{1}{\sqrt{2\pi}} \int_x^{\infty} e^{-t^2/2} \, dt$. Since statistics of ρ_j are independent of the index j when $j \neq m$, the sum operator in Eq. (6.20) can be replaced with the factor $M - 1$, and the inequality in P_e^{one} simplifies to

$$P_e^{one} \leq (M - 1) \int_{-\infty}^{\infty} Q(\rho_m \sqrt{N}) f_{\rho_m}(\rho_m) \, d\rho_m. \qquad (6.21)$$

6.2.2.1 Distribution of ρ_{ind}

If $\mathbf{W}_m$ and $\widehat{\mathbf{W}}_m$ have a zero covariance matrix, $\widehat{\mathbf{W}}_m$ carries no information about $\mathbf{W}_m$ due to the channel noise, and the normalized correlation ρ_{ind} between $\mathbf{W}_m$ and $\widehat{\mathbf{W}}_m$ is defined as

$$\rho_{ind} = \frac{\mathbf{W}_m^T \widehat{\mathbf{W}}_m}{\|\mathbf{W}_m\| \, \|\widehat{\mathbf{W}}_m\|} = \sum_{l=1}^{l=N} \frac{W_{m_l} \widehat{W}_{m_l}}{\|\mathbf{W}_m\| \, \|\widehat{\mathbf{W}}_m\|}. \qquad (6.22)$$

The random variable W_{m_l}, $1 \leq l \leq N$, has the variance $\|\mathbf{W}_m\|^2/N$ due to the *iid* assumption, where $\|\mathbf{W}_m\|^2$ is the power of $\mathbf{W}_m$. Similarly, the variance of $\widehat{W}_{m_l}$ is $\|\widehat{\mathbf{W}}_m\|^2/N$ independent of its pdf. Hence, the normalized random variables $W_{m_l}/\|\mathbf{W}_m\|$ and $\widehat{W}_{m_l}/\|\widehat{\mathbf{W}}_m\|$ are both zero-mean with variance $\frac{1}{N}$. The normalized correlation ρ_{ind} is a random variable with the mean $m_{\rho_{ind}}$ and the variance $\sigma_{\rho_{ind}}^2$ calculated as

$$m_{\rho_{ind}} = \sum_{l=1}^{l=N} E\left[\frac{W_{m_l}}{\|\mathbf{W}_m\|} \right] E\left[\frac{\widehat{W}_{m_l}}{\|\widehat{\mathbf{W}}_m\|} \right],$$

$$= 0, \qquad (6.23)$$

$$\sigma^2_{\rho_{ind}} = \sum_{l=1}^{l=N} Var\left[\frac{W_{m_l}}{\|\mathbf{W}_m\|}\right] Var\left[\frac{\widehat{W}_{m_l}}{\|\widehat{\mathbf{W}}_m\|}\right],$$

$$= N\frac{1}{N^2} = \frac{1}{N}. \tag{6.24}$$

The random variable ρ_{ind} has approximately Gaussian distribution, due to the central limit theorem, $\rho_{ind} \sim \mathcal{N}(0, \frac{1}{N})$.

Similarly, if $\mathbf{W}_m$ and $\mathbf{W}_j$ are independent *iid* random vectors, then $\widehat{\mathbf{W}}_m$ is also independent with $\mathbf{W}_j$. Consequently, the normalized correlation $\rho_{m,j} \sim \rho_{ind}$.

6.2.2.2   Distribution of ρ_{dep}

When $\mathbf{W}_m$ and $\widehat{\mathbf{W}}_m$ are dependent, a similar analysis can be performed. However, in this case, the samples W_{m_l} and $\widehat{W}_{m_l}$, $1 \le l \le N$, are somewhat correlated. The normalized correlation ρ_{dep}, defined between $\mathbf{W}_m$ and $\widehat{\mathbf{W}}_m$, is the normalized inner product of the two *iid* correlated random vectors, as given in Eq. (6.22).

For relatively small N, the embedding distortion P introduced into $\mathbf{C}$ with the use of optimal embedding parameters (computed for large N) becomes a random variable distributed around $P_E = \sigma^2_{X_n}$ with the variance $\frac{\sigma^2_P}{N}$ as discussed in Section 6.2. Based on the measured distortion P, the embedder has to adjust the processing distortion $\mathbf{X}_t$ by changing β or α in order to ensure an embedding distortion of P_E. Consequently, the effective noise level, $\mathbf{Z}_{eff} = \mathbf{Z} - \mathbf{X}_t$, at the detector changes, and the embedded signal $\mathbf{W}_m$ is distorted accordingly. The relation between the embedded binary watermark signal samples and the extracted samples is expressed in terms of Z_{eff} as in Eq. (5.30). The pdf of Z_{eff} for thresholding and distortion-compensation types of postprocessing is given in Eqs. (5.27) and (5.28) as a function of embedding parameters. Ultimately, the correlation coefficient ρ_{dep} between the dependent $\mathbf{W}_m$ and $\widehat{\mathbf{W}}_m$ can be calculated in terms of embedding parameters, N, and statistics of Z_{eff} and W.

It should be noted that a change in the embedding parameter β or α will induce a similar change in the value of the correlation coefficient as the parameter designates the amount of processing distortion applied. When N is not large enough, the embedding distortion P deviates from $P_E = \sigma_{X_n}^2$. This is reflected as a deviation of embedding parameters from their optimal values so that the adjusted β or α value yields $P = P_E$. Hence, the correlation of $\mathbf{W}_m$ and $\widehat{\mathbf{W}}_m$ is actually a random variable conditioned on P, $\rho_{dep}|P$, with the mean m_{ρ^*} and the variance $\sigma_{\rho^*}^2$. Its mean m_{ρ^*} is calculated as

$$
\begin{aligned}
m_{\rho^*} &= E\left[\frac{\mathbf{W}_m^T \widehat{\mathbf{W}}_m}{\|\mathbf{W}_m\| \, \|\widehat{\mathbf{W}}_m\|} \right] \\
&= \frac{E[W_m \widehat{W}_m]}{\sqrt{E[W_m^2]E[\widehat{W}_m^2]}} \\
&= \frac{R(1)}{\sqrt{R(2)}}
\end{aligned}
\tag{6.25}
$$

where $E[W_m^p \widehat{W}_m^p]$ is the pth joint moment of random variables W_m and $\widehat{W}_m$ and

$$
R(p) = 2 \sum_{i=0}^{i=\infty} \int_{\frac{i\Delta}{2}}^{\frac{(i+1)\Delta}{2}} \left(\left(\frac{(2i+1)\Delta}{4} - z_{eff} \right) (-1)^i \right)^p f_{Z_{eff}}(z_{eff}) \, dz_{eff}.
\tag{6.26}
$$

Similarly, the variance $\sigma_{\rho^*}^2$ of the random variable $\rho_{dep}|P$ is expressed as

$$
\sigma_{\rho^*}^2 = Var\left[\frac{\mathbf{W}_m^T \widehat{\mathbf{W}}_m}{\|\mathbf{W}_m\| \, \|\widehat{\mathbf{W}}_m\|} \right].
\tag{6.27}
$$

The details of the derivations for the Eqs. (6.25) and (6.27) are given in Appendix B.

The covariance matrix of the *iid* signal vector $\mathbf{W}_m$ and the extracted signal vector $\widehat{\mathbf{W}}_m$ is diagonal (i.e., $E[W_{m_l} \widehat{W}_{m_s}] = 0$, if $l \neq s$, $1 \leq l, s \leq N$). Therefore, the distribution of random variable $\rho_{dep}|P$ approximates Gaussian distribution, $\rho_{dep}|P \sim \mathcal{N}(m_{\rho^*}, \sigma_{\rho^*}^2)$, with mean and variance as given in Eqs. (6.25) and (6.27), respectively. The pdf of ρ_{dep} is therefore

$$f_{\rho_{dep}}(\rho_{dep}) = \int_{-\infty}^{\infty} f_{\rho_{dep}|P}(\rho_{dep}|P) f_P(P) \, dP \tag{6.28}$$

where $P \sim \mathcal{N}(\sigma_{X_n}^2, \frac{\sigma_P^2}{N})$.

6.2.3 Multiple Codebook Data Hiding Using the Maximum Correlation Criterion

In the multiple codebook data hiding method, the transmitted codeword, corresponding to a message, is expected to yield the highest detection statistics at the presumed noise level σ_Z^2. The embedder achieves this by searching for the transformation basis that yields less processing distortion than the others. This is achieved by choosing the maximum of the correlations $\tilde{\rho}_{m,m}^i$, $i = 1, \ldots, L$, that are measured between $\mathbf{W}_m$ embedded into L transformations of $\mathbf{C}$ and the corresponding extractions $\tilde{\mathbf{W}}_m^i$. However, due to channel noise $\mathbf{Z}$, the dependency between the embedded watermark signal and the extracted signal at the detector reduces. Therefore, the correlation $\tilde{\rho}_{m,m}^i$, between $\mathbf{W}_m$ and its extracted version from $\mathbf{Y}$, would be less than $\tilde{\rho}_{m,m}^i$ measured at the embedder. The correlation values $\tilde{\rho}_{m,m}^i$ and $\rho_{m,m}^i$ can be calculated from Eq. (5.30) for $\mathbf{Z}_{eff} = -\mathbf{X}_t$ and $\mathbf{Z}_{eff} = \mathbf{Z} - \mathbf{X}_t$, respectively. Ultimately, the transformation basis that yields the highest correlation at the embedder will also yield the highest correlation at the detector, $\arg_i \max \left(\tilde{\rho}_{m,m}^i \right) = \arg_i \max \left(\rho_{m,m}^i \right)$.

Let the maximum of $\rho_{m,m}^i$ be denoted by ρ_{max} with the pdf given as

$$\rho_{max} \sim \max \left(\rho_{m,m}^1, \ldots, \rho_{m,m}^L \right) \tag{6.29}$$

where $\rho_{m,m}^i$ are independent random variables with $\rho_{m,m}^i \sim \rho_{dep}$, Section 6.2.2.2. With multiple codebook hiding, then, detection errors are due to any of the normalized correlation values $\rho_{m,j}^i$, $j \neq m$, being greater than the correlation value ρ_{max}. Compared with the single codebook case, probability of error for multiple codebook hiding, P_e^{mul}, is expected to increase with the number of codebooks, as there are L times more normalized correlation values that can exceed ρ_{max}. On the other hand, since ρ_{max} is expected to have a higher mean than $\rho_{m,m}$, the probability of error for each comparison of the normalized correlations is reduced.

Assuming $\mathbb{T}_k$ is the transformation basis used for embedding in all cases, an event E_j^i that the detector will pick $\hat{m}$ instead of m is denoted as

$$E_j^i = \{p(\rho_{m,j}^i \geq \rho_{max})\}, \quad i = 1\dots,L, \quad j = 1,\dots,M \text{ and } j \neq m.$$
(6.30)

The event E^{mul} that the detector makes an error is

$$E^{mul} = \bigcup_{i=1}^{L} \bigcup_{j=1, j \neq m}^{M} E_j^i.$$
(6.31)

Hence, the probability of detecting a wrong message for multiple codebook hiding, P_e^{mul}, is obtained as

$$P_e^{mul} = \Pr\{E^{mul}\} \leq \sum_{i=1}^{L} \sum_{j=1, j \neq m}^{M} \Pr\{E_j^i\}.$$
(6.32)

The union bound on the probability of error can be rewritten using Eq. (6.30) as

$$P_e^{mul} \leq \sum_{i=1}^{L} \sum_{j=1, j \neq m}^{M} \Pr(\rho_{m,j}^i \geq \rho_{max}).$$
(6.33)

Comparing Eq. (6.17) with Eq. (6.33), one sees that the advantage of multiple codebook embedding over single codebook embedding is reflected in the statistics of $\rho_{m,m}$ and ρ_{max}.

The distribution of $\rho^i_{m,j}$, $1 \leq j \leq M$ and $1 \leq i \leq L$, can be generalized as

$$
\rho^i_{m,j} \sim
\begin{cases}
\mathcal{N}(0, \frac{1}{N}), & \text{if } i \neq k, \\
\mathcal{N}(0, \frac{1}{N}), & \text{if } i = k \text{ and } j \neq m, \\
\rho_{dep}, & \text{if } i = k \text{ and } j = m.
\end{cases}
\tag{6.34}
$$

The probability of error for multiple codebook hiding, Eq. (6.33), can be further rewritten using the preceding results as

$$
P^{mul}_e \leq \sum_{i=1}^{L} \sum_{j=1, j \neq m}^{M} \int_{-\infty}^{\infty} \int_{-\infty}^{\infty} f_{\rho^i_j}(\rho^i_j \geq \rho_{max}) f_{\rho_{max}}(\rho_{max}) \, d\rho^i_j \, d\rho_{max},
\tag{6.35}
$$

$$
\leq \sum_{i=1}^{L} \sum_{j=1, j \neq m}^{M} \int_{-\infty}^{\infty} \left(\int_{\rho_{max}}^{\infty} f_{\rho^i_j}(\rho^i_j) \, d\rho^i_j \right) f_{\rho_{max}}(\rho_{max}) \, d\rho_{max}
\tag{6.36}
$$

where the first subscript referring to the transmitted message m is dropped. Since the inner integral in Eq. (6.36) is the Gaussian Q-function and does not depend on the index j, Eq. (6.36) can be simplified to

$$
P^{mul}_e \leq L(M - 1) \int_{-\infty}^{\infty} Q(\rho_{max}\sqrt{N}) f_{\rho_{max}}(\rho_{max}) \, d\rho_{max}.
\tag{6.37}
$$

Note that for $L = 1$, P^{mul}_e given in Eq. (6.37) reduces to P^{one}_e in Eq. (6.21).

6.2.3.1   Distribution of $\rho^i_{m,j}$

The distribution of the random variables $\rho^i_{m,j}$ can be found based on the choice of i and j. When the detector assumes $i = k$, the transformations used for embedding and detection are the same. Then, the extracted signal

$\widehat{\mathbf{W}}_m^k$ is expressed as

$$\widehat{\mathbf{W}}_m^k = \mathcal{D}\left(\mathbb{T}_k\left(\mathbb{T}_k^T \mathcal{E}\left(\mathbb{T}_k\mathbf{S}, \mathbf{W}_m\right) + \mathbf{Z}\right)\right) \tag{6.38}$$
$$= \mathcal{D}\left(\mathcal{E}\left(\mathbb{T}_k\mathbf{C}, \mathbf{W}_m\right) + \mathbf{Z}'\right). \tag{6.39}$$

Since $\mathbf{Z}$ is assumed to be a white noise vector (*iid* Gaussian), a unitary transformation of it, $\mathbf{Z}' = \mathbb{T}_k\mathbf{Z}$, is also *iid* Gaussian with the same mean and variance. Therefore, the results of the analysis given in Sections 6.2.2.1 and 6.2.2.2 also apply to multiple codebook hiding. Consequently, the normalized correlation $\rho_{m,j}^k$, $1 \le j \le M$, is equivalent to random variables ρ_{dep} and ρ_{ind} in its statistics for $j = m$ and $j \ne m$, respectively.

   If there is a mismatch between the embedding and detection transformations, $i \ne k$, then $\widehat{\mathbf{W}}_m^i$ is obtained as

$$\widehat{\mathbf{W}}_m^i = \mathcal{D}\left(\mathbb{T}_i\left(\mathbb{T}_k^T \mathcal{E}\left(\mathbb{T}_k\mathbf{C}, \mathbf{W}_m\right) + \mathbf{Z}\right)\right) \tag{6.40}$$
$$= \mathcal{D}\left(\mathbb{T}_i\mathbb{T}_k^T \mathcal{E}\left(\mathbb{T}_k\mathbf{C}, \mathbf{W}_m\right) + \mathbf{Z}'\right) \tag{6.41}$$

where $\mathbf{Z}' = \mathbb{T}_i\mathbf{Z}$. In Eq. (6.41), $\widehat{\mathbf{W}}_m^i$ is related to $\mathbf{W}_m$ through the transformation $\mathbb{T}_i$ followed by a nonlinear detection (see Section 5.2). For properly selected transform bases, $E[\|\mathbb{T}_i\mathbf{C} - \mathbb{T}_k\mathbf{C}\|]$ is maximized. An extraction from $\mathbb{T}_i$ transformation of the received signal does not provide any meaningful information about $\mathbf{W}_m$ because the embedding transformation was $\mathbb{T}_k$. Consequently, the binary distributed $\mathbf{W}_m$ with values in $\{-\frac{\Delta}{4}, \frac{\Delta}{4}\}$, is extracted, $\widehat{\mathbf{W}}_m^i$, as a uniformly distributed sample sequence in the range $[-\frac{\Delta}{4}, \frac{\Delta}{4}]$ which is independent of $\mathbf{W}_m$. Therefore, the normalized correlation $\rho_{m,j}^i$, $i \ne k$ and $\forall j$, has the same statistics as the random variable ρ_{ind}, $\rho_{m,j}^i \sim \mathcal{N}(0, \frac{1}{N})$.

6.2.3.2   Distribution of ρ_{max}

The random variable ρ_{max} is the maximum of L random variables, Eq. (6.29), that are all distributed according to pdf of random variable ρ_{dep}. The distribution of ρ_{max}, for any finite L, can be expressed in terms of the distribution function of ρ_{dep} as

$$F_{\rho_{max}}(\rho_{max}) = F^L_{\rho_{dep}}(\rho_{max}) \tag{6.42}$$

where $F_X(x) = \int_{-\infty}^{x} f_X(x)\,dx$ and the superscript L refers to the Lth-order power of the distribution function $F_{\rho_{dep}}(\rho_{max})$. Correspondingly, the pdf of ρ_{max} is found as $f_{\rho_{max}}(\rho_{max}) = LF^{L-1}_{\rho_{dep}}(\rho_{max})f_{\rho_{dep}}(\rho_{max})$.

6.2.4 Single Codebook Hiding Using the Minimum Distance Criterion

Considering the minimum distance criterion for the single codebook hiding case, a detection error is the result of $\widehat{\mathbf{W}}_m$ having the smallest distance with any of $\{\mathbf{W}_1, \ldots, \mathbf{W}_M\}$ other than $\mathbf{W}_m$. Hence, the upper bound on the probability of detection error, P_e^{one}, can be expressed similarly to Section 6.2.2, Eqs. (6.14)–(6.17), as

$$P_e^{one} \leq \sum_{j=1, j\neq m}^{M} p(d_{m,j} \leq d_{m,m}). \tag{6.43}$$

As will be shown in the following sections, the statistics of the random variables $d_{m,j}$ and $d_{m,m}$ in Eq. (6.43) are, respectively, the same as those of d_{ind} and d_{dep}. Consequently, the pdf of random variable $d_{m,j}$, $1 \leq m,j \leq M$, can be expressed as

$$d_{m,j} \sim \begin{cases} \mathcal{N}\left(\dfrac{\Delta^2}{12}, \dfrac{\Delta^4}{N180}\right), & \text{if } m \neq j \\ d_{dep}, & \text{if } m = j. \end{cases} \tag{6.44}$$

Assuming that m is the index of the transmitted message for the generic case, Eq. (6.43) can be rewritten using Eq. (6.44) as

$$P_e^{one} \leq \sum_{j=1, j\neq m}^{M} \int_{-\infty}^{\infty} \int_{-\infty}^{\infty} f_{d_j}(d_j \leq d_m)f_{d_m}(d_m)\,dd_j\,dd_m \tag{6.45}$$

$$\leq \sum_{j=1,\, j\neq m}^{M} \int_{-\infty}^{\infty} \left(\int_{-\infty}^{d_m} f_{d_j}(d_j)\, dd_j \right) f_{d_m}(d_m)\, dd_m \tag{6.46}$$

$$\leq (M-1) \int_{-\infty}^{\infty} F_{d_j}(d_m) f_{d_m}(d_m)\, dd_m \tag{6.47}$$

where $F_{d_j}(d_j)$ is the probability distribution function of the random variable d_j.

6.2.4.1 Distribution of d_{ind}

When $\mathbf{W}_m$ and $\widehat{\mathbf{W}}_m$ have a zero covariance matrix, the distance d_{ind} between the *iid* $\mathbf{W_m}$ and $\widehat{\mathbf{W}}_m$ can be defined as

$$d_{ind} = \frac{1}{N} \|\mathbf{W}_m - \widehat{\mathbf{W}}_m\|^2$$

$$= \frac{1}{N} (\mathbf{W}_m - \widehat{\mathbf{W}}_m)^T (\mathbf{W}_m - \widehat{\mathbf{W}}_m)$$

$$= \frac{1}{N} \sum_{l=1}^{l=N} (W_{m_l}^2 + \widehat{W}_{m_l}^2 - 2W\widehat{W}_{m_l}). \tag{6.48}$$

Introducing the random variable $\lambda = W^2 + \widehat{W}^2 - 2W\widehat{W}$, such that $d_{ind} = \frac{1}{N}\sum_{l=1}^{l=N} \lambda_{m_l}$, the statistics of random variable d_{ind} can be computed in terms of the statistics of λ. The mean and variance of λ are, respectively, derived in Appendix B as

$$m_\lambda = \frac{\Delta^2}{12}, \tag{6.49}$$

$$\sigma_\lambda^2 = \frac{\Delta^4}{180}. \tag{6.50}$$

Therefore,

$$m_{d_{ind}} = E\left[\frac{1}{N} \sum_{j=1}^{j=N} \lambda_{m_l} \right]$$

$$= \frac{1}{N} N m_\lambda = \frac{\Delta^2}{12}, \tag{6.51}$$

$$\sigma_{d_{ind}}^2 = Var\left[\frac{1}{N}\sum_{j=1}^{j=N}\lambda\right],$$

$$= \frac{1}{N^2}N\sigma_\lambda^2 = \frac{1}{N}\frac{\Delta^4}{180}. \tag{6.52}$$

As both $\mathbf{W}_m$ and $\widehat{\mathbf{W}}_m$ are *iid*, the distribution of d_{ind} approximates Gaussian, $d_{ind} \sim \mathcal{N}(\frac{\Delta^2}{12}, \frac{\Delta^4}{N180})$. Similarly, the distance $d_{i,j}$ measured between the extracted signal $\widehat{\mathbf{W}}_i$ and the watermark signal $\mathbf{W}_j$ is equivalent to d_{ind} in its statistics when $\mathbf{W}_i$ and $\mathbf{W}_j$ are mutually independent *iid* random vectors.

6.2.4.2   Distribution of d_{dep}

When $\mathbf{W}_m$ and $\widehat{\mathbf{W}}_m$ have a diagonal covariance matrix, an analysis similar to the one given in Section 6.2.2.2 is performed. The distance d_{dep} is the mean squared difference of the *iid* correlated random vectors $\mathbf{W}_m$ and $\widehat{\mathbf{W}}_m$, as defined in Eq. (6.48). Given that optimal embedding parameters yield an embedding distortion of P, the distance between $\widehat{\mathbf{W}}_m$ and $\mathbf{W}_m$ can be expressed as a random variable conditioned on P. The mean m_{d*} and the variance σ_{d*}^2 of $d_{dep}|P$ can be calculated in terms of the statistics of λ_{m_l} as

$$m_{d*} = E\left[\frac{1}{N}\sum_{l=1}^{N}\lambda_{m_l}\right]$$

$$= \left(\frac{\Delta}{4}\right)^2 - 2\frac{\Delta}{4}R(1) + R(2), \tag{6.53}$$

$$\sigma_{d*}^2 = Var\left[\frac{1}{N}\sum_{l=1}^{N}\lambda_{m_l}\right]$$

$$= \left(\frac{\Delta}{4}\right)^4 - 4\left(\frac{\Delta}{4}\right)^3 R(1) + \frac{1}{N}\left(R(4) + 6\left(\frac{\Delta}{4}\right)^2 R(2) - \Delta R(3)\right)$$

$$\times \frac{N-1}{N}\left(2\left(\frac{\Delta}{4}\right)^2 R(2) + 2\left(\frac{\Delta}{4}\right)^2 R(1)^2 + 4\left(\frac{\Delta}{4}\right)^2 R(1)^2 - \Delta R(1)R(2)\right.$$

$$\left. + R(2)^2\right) - m_{d*}^2 \tag{6.54}$$

where $R(p)$ is as given in Eq. (6.26). Derivation details for Eqs. (6.53) and (6.54) are given in Appendix B, Eqs. (B.12)–(B.13). The distribution of $d_{dep}|P$ also converges to a Gaussian distribution, $d_{dep}|P \sim \mathcal{N}(m_{d*}, \sigma_{d*}^2)$. The pdf of random variable d_{dep} is calculated as

$$f_{d_{dep}}(\rho_{dep}) = \int_{-\infty}^{\infty} f_{d_{dep}|P}(d_{dep}|P) f_P(P)\, dP \tag{6.55}$$

where $P \sim \mathcal{N}(\sigma_{X_n}^2, \frac{\sigma_P^2}{N})$.

6.2.5 Multiple Codebook Hiding Using the Minimum Distance Criterion

In this version of the method, the embedder selects the transformation basis by choosing the minimum of the distances $\tilde{d}_{m,m}^i$, $i = 1, \ldots, L$ computed between $\mathbf{W}_m$ and $\tilde{\mathbf{W}}_m^i$ for each transformation of $\mathbf{C}$. At the detector, on the other hand, the distance between the embedded and the extracted signals is measured as $d_{m,m}^i$, $1 \le i \le L$. The degradation in the measured distance from $\tilde{d}_{m,m}^i$ to $d_{m,m}^i$ is due to the channel noise $\mathbf{Z}$ as discussed in Section 6.2.3. However, the transformation basis that yields the minimum distance at the embedder will yield the minimum distance at the detector, $\arg_i \min(\tilde{d}_{m,m}^i) = \arg_i \min\left(d_{m,m}^i\right)$. Defining the minimum of $d_{m,m}^i$ as d_{min}, its pdf is given as

$$d_{min} \sim \min\left(d_{m,m}^1, \ldots, d_{m,m}^L\right) \tag{6.56}$$

where $d_{m,m}^i$ are independent random variables with $d_{m,m}^i \sim d_{dep}$ (see Section 6.2.4.2). Consequently, a detection error occurs if any of the distance values $d_{m,j}^i$, $1 \le j \le M$, $j \ne m$, and $1 \le i \le L$ are smaller than

d_{min}. Compared with the single codebook case, similar to Section 6.2.3, probability of error is expected to increase with respect to the number of codebooks because there are L times more distance values that may be smaller than d_{min}, whereas d_{min} has a lower mean than $d_{m,m}$, which will reduce the probability of error. The union bound on the probability of error for multiple codebook hiding, P_e^{mul}, is found to be similar to Eqs. (6.30)–(6.33) as

$$P_e^{mul} \le \sum_{i=1}^{L} \sum_{j=1, j \neq m}^{M} \Pr(d_{m,j}^i \le d_{min}). \tag{6.57}$$

The advantage of multiple codebook hiding stems from the difference in the distributions of the random variables $d_{m,m}$ and d_{min} in Eqs. (6.43) and (6.57), respectively. The distribution of $d_{m,j}^i$, $1 \le j \le M$ and $1 \le i \le L$, can be generalized as

$$d_{m,j}^i \sim \begin{cases} \mathcal{N}(\frac{\Delta^2}{12}, \frac{\Delta^4}{N180}), & \text{if } i \neq k, \\ \mathcal{N}(\frac{\Delta^2}{12}, \frac{\Delta^4}{N180}), & \text{if } i = k \text{ and } j \neq m, \\ d_{dep}, & \text{if } i = k \text{ and } j = m. \end{cases} \tag{6.58}$$

The bound on the probability of error given in Eq. (6.57) can be rewritten using these results (by dropping the first subscript referring to the transmitted message m) as

$$P_e^{mul} \le \sum_{i=1}^{L} \sum_{j=1, j \neq m}^{M} \int_{-\infty}^{\infty} \int_{-\infty}^{\infty} f_{d_j^i}(d_j^i \le d_{min}) f_{d_{min}}(d_{min}) \, dd_j^i \, dd_{min}, \tag{6.59}$$

$$\le \sum_{i=1}^{L} \sum_{j=1, j \neq m}^{M} \int_{-\infty}^{\infty} \left(\int_{-\infty}^{d_{min}} f_{d_j^i}(d_j^i) \, dd_j^i \right) f_{d_{min}}(d_{min}) \, dd_{min}, \tag{6.60}$$

$$\le L(M-1) \int_{-\infty}^{\infty} F_{d_j^i}(d_{min}) f_{d_{min}}(d_{min}) \, dd_{min} \tag{6.61}$$

where $d_j^i \sim \mathcal{N}(m_{d_{ind}}, \sigma^2_{d_{ind}})$.

6.2.5.1 Distribution of $d_{m,j}^i$

The distribution of the random variables $d_{m,j}^i$ can be found based on the choice of i and j, as in Section 6.2.3. When the detector assumes that $i = k$, the transformations used for embedding and detection are the same. The detected watermark signal $\widehat{\mathbf{W}}_{m,j}^k$ can be expressed as in Eq. (6.39). Thus, the analysis given for the single codebook case also applies to the multiple codebook case. The distance between the $\mathbf{W}_m$ and $\widehat{\mathbf{W}}_{m,j}^k$, $d_{m,j}^k$ for $1 \leq j \leq M$ and $j \neq m$, has the same statistics as the random variable d_{ind}, $d_{m,j}^k \sim \mathcal{N}\left(\frac{\Delta^2}{12}, \frac{\Delta^4}{N180}\right)$. In the same manner, $d_{m,m}^k$, $j = m$, has the same statistics as the random variable d_{dep}, $d_{m,m}^k \sim d_{dep}$.

If there is a mismatch between the embedding and detection transformations such that $i \neq k$, then $\widehat{\mathbf{W}}_{m,j}^k$ is obtained as in Eq. (6.41). Due to the transformation $\mathbb{T}_i$, $i \neq k$, and the nonlinear detection that follows it, $\widehat{\mathbf{W}}_{m,j}^k$ becomes independent of $\mathbf{W}_m$. Therefore, the mean squared distance, $d_{m,j}^i$ for $i \neq k$, is equivalent to the random variable d_{ind} in its statistics, $d_{m,j}^i \sim \mathcal{N}(\frac{\Delta^2}{12}, \frac{\Delta^4}{N180})$.

6.2.5.2 Distribution of d_{min}

Since d_{min} is the minimum of L independent random variables, Eq. (6.56), distributed according to $F_{d_{dep}}(d_{dep})$, the probability distribution function of d_{min} is found as

$$F_{d_{min}}(d_{min}) = 1 - \left(1 - F_{d_{dep}}(d_{min})\right)^L. \tag{6.62}$$

The pdf of random variable d_{min} is therefore

$$f_{d_{min}}(d_{min}) = L\left(1 - F_{d_{dep}}(d_{min})\right)^{L-1} f_{d_{dep}}(d_{min}). \tag{6.63}$$

6.2.6 Comparisons

The robustness measure used to compare multiple codebook hiding with single codebook hiding is defined in terms of the ratio between the embedding distortion power and the channel noise power, WNR$= \frac{P_E}{\sigma_Z^2}$. Figures

6-11–6-13 and 6-14–6-16 display the union bound on the probability of error for the thresholding type of postprocessing using both criteria. The curves are obtained by numerically solving Eqs. (6.37) and (6.61) at different WNRs and for various numbers of codebooks and codebook sizes $M \times N$. Corresponding results for the distortion-compensation type of postprocessing are similarly displayed in Figs. 6-17–6-19 and 6-20–6-22. In all cases, as the number of codebooks increases, the bound on the probability

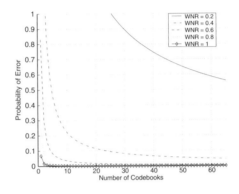

Figure 6-11 Probability of error performance for multiple codebook hiding based on maximum correlation criterion and thresholding type of processing for $M = 100$ and $N = 50$.

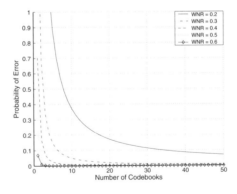

Figure 6-12 Probability of error performance for multiple codebook hiding based on maximum correlation criterion and thresholding type of processing for $M = 200$ and $N = 100$.

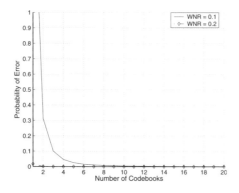

Figure 6-13   Probability of error performance for multiple codebook hiding based on maximum correlation criterion and thresholding type of processing for $M = 1000$ and $N = 500$.

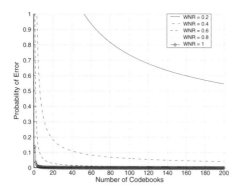

Figure 6-14   Probability of error performance for multiple codebook hiding based on minimum distance criterion and thresholding type of processing for $M = 100$ and $N = 50$.

of error decreases exponentially. On the other hand, the probability of error for single codebook hiding also decreases with the increasing signal size N. Consequently, fewer codebooks are required to further improve the performance. Results show that for WNR ≥ 1 and WNR ≥ 0.2 (equivalently in logarithmic scale WNR ≥ 0 dB and WNR ≥ -7 dB) the use of multiple codebooks is not necessary if $N \simeq 100$ and $N \simeq 500$,

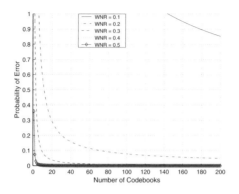

Figure 6-15 Probability of error performance for multiple codebook hiding based on minimum distance criterion and thresholding type of processing for $M = 200$ and $N = 100$.

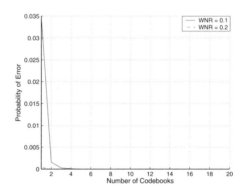

Figure 6-16 Probability of error performance for multiple codebook hiding based on minimum distance criterion and thresholding type of processing for $M = 1000$ and $N = 500$.

respectively. Intuitively, this is due to increasing confidence in the detection with the increasing N. With reference to the analyses in Sections 6.2.3 and 6.2.5, as $m_{\rho_{dep}}$ increases and $\sigma^2_{\rho_{dep}}$ decreases, the maximum of the ensemble of random variables $\tilde{\rho}^1_{m,m}, \ldots, \tilde{\rho}^L_{m,m}$ is less likely to differ from the rest. Respectively, as $m_{d_{dep}}$ decreases, the minimum of $\tilde{d}^1_{m,m}, \ldots, \tilde{d}^L_{m,m}$

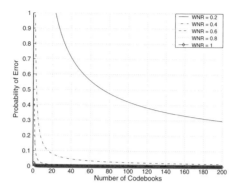

Figure 6-17   Probability of error performance for multiple codebook hiding based on maximum correlation criterion and distortion-compensation type of processing for $M = 100$ and $N = 50$.

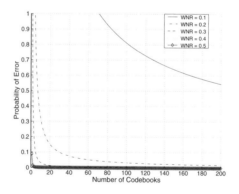

Figure 6-18   Probability of error performance for multiple codebook hiding based on maximum correlation criterion and distortion-compensation type of processing for $M = 200$ and $N = 100$.

will not differ significantly from any of the other measured distances. Consequently, all codebooks become almost equally favorable.

In the multiple codebook data hiding method, since the detector forces the extracted signal to match one of the watermark signals, one concern is the probability of a false positive (false alarm). This is the probability of

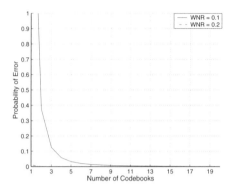

Figure 6-19   Probability of error performance for multiple codebook hiding based on maximum correlation criterion and distortion-compensation type of processing for $M = 1000$ and $N = 500$.

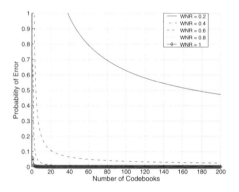

Figure 6-20   Probability of error performance for multiple codebook hiding based on minimum distance criterion and distortion-compensation type of processing for $M = 100$ and $N = 50$.

detecting a message when no message is embedded, and it can be derived based on the results of analysis given in Sections 6.2.2 and 6.2.3. Under the assumption that the host signal is distributed uniformly in each quantization interval ($\sigma_C^2 \gg \Delta$), the extracted signal $\widehat{\mathbf{W}}_{null}$ is *iid* uniformly distributed in $[-\frac{\Delta}{4}, \frac{\Delta}{4}]$ and uncorrelated with any of the watermark signals. As a result, the normalized correlation $\rho_{null,\,j}$ or the squared error distance $d_{null,\,j}$

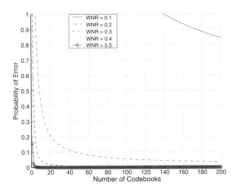

Figure 6-21 Probability of error performance for multiple codebook hiding based on minimum distance criterion and distortion-compensation type of processing for $M = 200$ and $N = 100$.

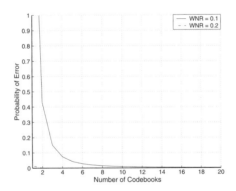

Figure 6-22 Probability of error performance for multiple codebook hiding based on minimum distance criterion and distortion-compensation type of processing for $M = 1000$ and $N = 500$.

between $\widehat{\mathbf{W}}_{null}$ and $\mathbf{W}_j$, $1 \leq j \leq M$, is distributed as $\mathcal{N}(0, \frac{1}{N})$ *irrespective of the channel noise level.*

For single codebook hiding, a false positive occurs when $\rho_{null,j}$ is greater or $d_{null,j}$ is smaller than a preset threshold. Using maximum correlation criterion, the threshold is set based on the statistics of ρ_{dep}, which is the normalized correlation between an embedded watermark signal and its extracted version, so that the embedded message can be distinguished from the rest at a constant false-alarm rate. Respectively, using minimum distance criterion, the threshold is determined based on the statistics of d_{dep}.

With multiple codebook hiding, where extractions are made from unitary transformations of the received signal, the extracted signals $\widehat{\mathbf{W}}^i_{null}$, $1 \leq i \leq L$, have the same statistics as $\widehat{\mathbf{W}}_{null}$. Consequently, the correlation $\rho^i_{null,j}$ and the distance $d^i_{null,j}$, computed between $\widehat{\mathbf{W}}^i_{null}$ and $\mathbf{W}_j$, have the same statistics as $\rho_{null,j}$ and $d_{null,j}$, respectively. Correspondingly, the probability of a false positive is due to $\rho^i_{null,j}$ being greater or $d^i_{null,j}$ being smaller than the preset threshold. Considering a fixed threshold for message detection, the false-alarm rate within multiple codebook hiding increases with a factor of L compared with single codebook hiding (as there are so many comparisons that may yield a false positive). However, noting that the use of multiple codebooks enables embedding a watermark signal with less processing distortion, the correlation and distance properties of the extracted signal are improved. Therefore, using the maximum correlation criterion, one can afford to increase the threshold in accordance with the statistics of ρ_{max}. Alternately, using the minimum distance criterion, the threshold can be decreased depending on the statistics of d_{min}.

The numerical solutions of Eq. (6.37) indicate that the increase in the P_e^{mul} by the factor of L, compared with P_e^{one}, is compensated by the embedder's ability to better adapt the codeword to the host signal, as a result of which detection statistics are improved from those of ρ_{dep} to ρ_{max}. Similarly, the linear increase in the false alarm rate with the number of codebooks can be compensated by an exponential decrease through proper selection of the threshold, which relies on the statistics of ρ_{max} rather than of ρ_{dep}. A similar reasoning based on the solution of Eq. (6.61) is valid for the minimum distance criterion due to the improvement in distance properties from d_{dep} to d_{min}.

A complete comparison of multiple codebook hiding and single codebook hiding schemes would involve calculating the actual probability of errors (not the union bound), which would be extremely difficult. However, the analytical results indicate that, as in Eqs. (6.37) and (6.61), the upper bound on the probability of error decreases exponentially for the multiple codebook data hiding scheme. Therefore, schemes employing multiple codebooks, rather than a single codebook, will perform better when N is limited.

6.2.7 Implementation and Simulation Results

Optimum codeword selection in multiple codebook hiding depends on designing the set of transform bases $\mathbb{T}_1, \ldots, \mathbb{T}_L$ properly (i.e., they should be able to generate maximally separated transformations of the host signal; see Eq. (6.13)). One intuitive way of picking such a set of transform bases is by choosing them among rotation matrices so that each transformation of the host signal is a rotated version of the others. The multiple codebook data hiding method is implemented by designing the transformation bases using Givens rotations [66]. Givens rotations provide orthogonal transformations in $\mathfrak{R}^N$ that can be employed to rotate a given vector with a chosen angle.

A particular transform basis $\mathbb{T}_k$ is obtained by the consecutive multiplication of $\frac{N(N-1)}{2}$ number of orthogonal matrices, all with determinant 1 so that the resulting $\mathbb{T}_k$ is unitary. Each orthogonal matrix is derived from the identity matrix by introducing $\cos \theta_k$ terms at (i, i) and (j, j) locations along with $\sin \theta_k$ and $-\sin \theta_k$ terms at (i, j) and (j, i) locations in order to rotate the (i, j) coordinate plane with the designated angle θ_k. The rotation angles θ_k, $k = 1, \ldots, L$, are chosen by uniformly sampling 2π, $\theta_k = (k - 1) \frac{2\pi}{L}$.

By setting the signal size to N and the number of messages to M, the size of the codebooks utilized by the embedder is fixed to $M \times N$. The watermark signals that are embedded into the host signal are generated using a Hadamard transform matrix due to its simplicity. The Hadamard transform matrix of size $N \times N$ and its negated version are combined into a $2N \times N$ binary valued matrix. Every row of the combined matrix is indexed from 1 to $M = 2N$, scaled by $\frac{\Delta}{4}$ for maximum separation, and assigned to the watermark signal $\mathbf{W}_j$, $1 \leq j \leq M$, such that $E[\mathbf{W}_i^T \mathbf{W}_j] = 0$, $i \neq j$ and $i \neq j + N$. The host signal and channel noise are *iid* zero-mean Gaussian vectors with $\sigma_C^2 \gg P_E$, σ_Z^2. Prior to embedding, the permitted embedding distortion P_E is fixed, and the optimal values for the embedding parameter Δ are derived for the considered WNRs. The Δ values are also revealed to the detector. The parameters β and α, however, are properly adjusted for each embedding in order to ensure an embedding distortion of P_E and are not known to the detector. The simulations are done for a different number of transformations L and signal sizes N by embedding and detecting randomly chosen message indices.

Multiple codebook hiding is implemented in the type III scheme based on thresholding and distortion-compensation types of postprocessing using both maximum correlation and minimum distance criteria. Message embedding and detection with up to 25 codebooks is performed considering codebook sizes of 64×32, 128×64, and 256×128 and the WNR range of 0. 1 to 1. Figures 6-23 and 6-24 display the probability of success results obtained, respectively, for $L = 1, 3$ and $L = 1, 4$ with varying

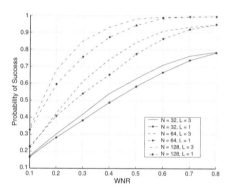

Figure 6-23   Probability of success performance for three-codebook hiding based on thresholding processing and maximum correlation criterion for various watermark signal sizes of $N = 32$, $N = 64$, and $N = 128$.

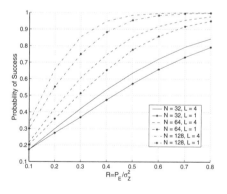

Figure 6-24   Probability of success performance for four-codebook hiding based on thresholding processing and minimum distance criterion for various watermark signal sizes of $N = 32$, $N = 64$, and $N = 128$.

N values where the postprocessing is thresholding. The increase in the embedding signal size N, at a fixed number of codebooks, improves the detection statistics because normalized correlation and mean squared distance give more reliable results with the larger signal sizes. On the other hand, Figs. 6-25 and 6-26 display the performances for a thresholding

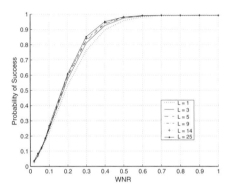

Figure 6-25 Probability of success performance for multiple codebook hiding based on thresholding type of processing and maximum correlation criterion for L = 1,3,5,9,14,25 and N = 128.

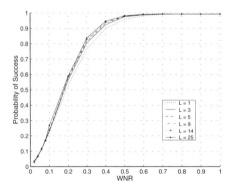

Figure 6-26 Probability of success performance for multiple codebook hiding based on thresholding type of processing and minimum distance criterion for L = 1,3,5,9,14,25 and N = 128.

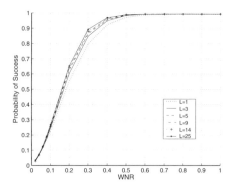

Figure 6-27 Probability of success performance for multiple codebook hiding based on distortion-compensation type of processing and maximum correlation criterion for L = 1,3,5,9,14,25 and N = 128.

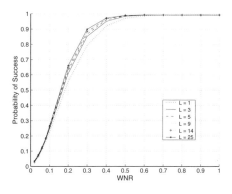

Figure 6-28 Probability of success performance for multiple codebook hiding based on distortion-compensation type of processing using minimum distance criterion for L = 1,3,5,9,14,25 and N = 128.

type of postprocessing when $N = 128$ and $L = 1, 3, 5, 9, 14, 25$ using the two criteria. Corresponding results for the distortion-compensation type of processing are displayed in Figs. 6-27 and 6-28 for both criteria. It is observed from these performance simulations that the multiple codebook data hiding method has superior performance vs the corresponding single codebook method at the same N.

The computational complexity of the proposed method depends on the number of codebooks employed. Multiple codebook embedding, when compared with single codebook embedding, requires the embedding of the watermark signal into transformations of the host signal and a comparison based on the resulting signals in order to select the transformation basis. On the other hand, at the detector, extraction should be repeated for each transformation basis. Therefore, the computational complexity increases almost linearly with the number of codebooks (see Fig. 6-10).

Major Design Issues

Though the material in the previous chapters has addressed the basis for a framework of optimal data hiding systems, there are many other aspects which need to be addressed to translate the theoretical framework to practical and usable designs. This chapter addresses some issues in the design of such systems.

In the previous chapters, we argued that the design of data hiding systems can be loosely divided into two parts—a conventional signaling scheme, which addresses the issue of mapping information bits to the watermark signal $\mathbf{W}$, and an embedder/detector $\mathcal{E}, \mathcal{D}$, which "mixes" the watermark signal and the cover signal to yield the stego signal. The first section of this chapter addresses the design of the former, viz., the pair $(\mathcal{W}, \mathcal{W}^{-1})$.

Further, the analysis of the data hiding systems presented in earlier chapters assumed "synchronous" communications. In the context of data hiding, this implies a shared knowledge of spatial or time coordinates between the embedder and the extractor. In practice, there are many disturbances in the channel that may result in loss of this synchronization. For example, if the cover signal were an image, resizing or cropping the stego image would result in loss of synchronization. Section 7.2 address this issue of synchronization. However, at the risk of losing some generality, we consider the issue of synchronization only for data hiding in images.

Another important issue to be addressed is that of ensuring perceptual transparency of the distortion introduced due to data hiding. The analysis in the previous chapters has measured perceptual transparency using mean square distortion of the watermark signal power. However, this measure is not always useful in practice. Section 7.3 briefly discusses issues involved in placing constraints on the distortion introduced to satisfy perceptual transparency.

7.1   DFT-Based Signaling

7.1.1   Conventional Signaling

The conventional signaling part, viz., the pair $(\mathcal{W}, \mathcal{W}^{-1})$, addresses the problem of mapping a K-length bit sequence b to a possibly real-valued sequence $\mathbf{W}$ of length N, where $N \gg K$. As a simple approach, we have

$$\mathbf{W} = [W_1 \, W_2 \, \cdots \, W_K] \tag{7.1}$$

where $W_i = \text{sign}(b(i))\theta, i = 1, \ldots, K$, and θ is random vector (obtained from a random seed or the private key $\mathcal{K}$) of length N/K. On the other hand, we could generate 2^K sequences $\mathbf{W}_i, i = 1, \ldots, 2^K$ of length N, such that the sequences $\mathbf{W}_k$ are *maximally separable*. Geometrically, the sequences $\mathbf{W}_k$ can be represented by a set of 2^K points in an N-dimensional hypersphere. In other words, the minimum distance between any two of 2^K points should be as high as possible, under the given constraint of the hypersphere radius. The binary sequence $[b_1 \, b_2 \, \cdots \, b_K]$ can be interpreted as a decimal number between 0 and $2^K - 1$. For instance, transmit a particular sequence of bits whose decimal equivalent is d, we choose $\mathbf{W} = \mathbf{W}_d$.

Detection of the hidden bit sequence, or equivalently the number d, can be accomplished as $\tilde{d} = \arg\max_{i=0\cdots2^K-1}\langle\widehat{\mathbf{W}}, \mathbf{W}_i\rangle$.

While it is assured that the latter scheme will approach the *channel capacity* closer than the former, in practice, implementation of the second scheme may be prohibitively expensive, especially for large K and/or N.

A reasonable compromise might be to choose an alphabet size between 2 of the former (bit-by-bit signaling) technique and 2^K of the latter. For example, if the alphabet size is chosen as $2^{K/k}$, then a single member of the alphabet is detected from each of the k sequences of length N/k.

A signaling method based on a fast Fourier transform (FFT) proposed in the next section offers an efficient way to increase the alphabet size used for signaling, while keeping the computational complexity at manageable levels. Furthermore, the maximally separable signal constellation itself is generated from random seeds.

7.1.2 FFT-Based Signaling

In the FFT-based signaling technique, the maximally separable sequences are constrained to be orthogonal. Let $\mathbf{W}_k \in \Re^{L_k}, L_k = 2^{p_k-1}$. Maximally separable signature sequences $\mathbf{W}_k^l, l = 1, \ldots, 2^{p_k}$, corresponding to p_k bits, are obtained as L_k orthogonal sequences and their negatives. *Random signature spaces* are generated from a seed. This is achieved by constraining the signatures to be *cyclic all-pass sequences*.

7.1.2.1 Cyclic All-Pass Sequences

Let $\boldsymbol{h} \in \Re^N$ and $\boldsymbol{H} = \mathcal{F}(\boldsymbol{h})$, where $\mathcal{F}(\cdot)$ stands for the DFT. Further, let $\boldsymbol{h}$ be such that

$$|H(n)| = 1 \quad \text{for } n = 0, 1, \ldots, N-1. \tag{7.2}$$

Hence

$$(\boldsymbol{H} \cdot \boldsymbol{H}^*) = [1, 1, \ldots, 1]. \tag{7.3}$$

Taking the inverse DFT (IDFT) of both sides of Eq. (7.3), we get

$$\mathcal{F}^{-1}(\boldsymbol{H}.\boldsymbol{H}^*) = [1, 0, 0, \ldots, 0]. \tag{7.4}$$

As $\mathcal{F}^{-1}(\boldsymbol{H} \cdot \boldsymbol{H}^*)$ is the *circular autocorrelation* of the vector $\boldsymbol{h}$, it follows that all circular shifts of $\boldsymbol{h}$ are mutually orthogonal [67]. As the phases $\phi_n, n = 0, 1, \ldots, N-1$ of the elements of $\boldsymbol{H}$ can be arbitrary, we have infinitely many choices for the vector $\boldsymbol{h}$ with mutually orthogonal circular shifts. For real $\boldsymbol{h}$ we have $N/2 - 1$ phase values, which can be arbitrarily chosen. Thus a pseudo-random all-pass sequence of length N can be generated from a pseudo-random (uniformly distributed between π and $-\pi$) sequence of length $N/2 - 1$. If

$$\phi_k = \begin{cases} 0 \text{ or } \pi & k = 0, k = N/2 \\ \theta_k & k = 0, \ldots, N/2 - 1 \\ -\theta_{N-k} & k = N/2 + 1, \ldots, N - 1 \end{cases}$$

$$H(k) = \cos(\phi_k) + \mathrm{i}\sin(\phi_k), \quad k = 0, \ldots, N - 1, \tag{7.5}$$

where $\theta_k, k = 1, \ldots, N/2 - 1$ are randomly distributed between π and $-\pi$, $\mathrm{i} = \sqrt{-1}$, then $\boldsymbol{h} = \mathcal{F}^{-1}(\boldsymbol{H})$ is a cyclic all-pass sequence.

Alternately, a pseudo-random binary sequence is generated from a seed. Then, the *unique* all-pass sequence "closest" (in the mean square sense) to the binary sequence is obtained (this guarantees that the signature energy will not be concentrated in a few coefficients).

Let $\boldsymbol{f} = [\, f(0)\, f(1) \cdots f(N - 1)]$ be a random binary sequence. We need to find the all-pass sequence that is closest to $\boldsymbol{f}$. In other words, we need to find the vector $\boldsymbol{h} = [h(0)\, h(1) \cdots h(N - 1)]^T$ that minimizes the error ε defined as

$$\varepsilon = \sum_{n=0}^{N-1} |h(n) - f(n)|^2 \tag{7.6}$$

subject to the constraint that $\boldsymbol{h}$ is a cyclic all-pass sequence. Since the DFT of a (cyclic) all-pass sequence can be written as $\boldsymbol{H} = [e^{j\phi_0}\, e^{j\phi_1} \cdots e^{j\phi_{N-1}}]$, let

$$h(n) = \sum_{k=0}^{N-1} e^{j(\frac{2\pi kn}{N} + \phi_k)} \qquad f(n) = \sum_{k=0}^{N-1} a_k e^{j(\frac{2\pi kn}{N} + \theta_k)}$$

for $n = 0, \ldots, N - 1$. It can be easily shown (see Appendix C) that the error ε is given by

$$\varepsilon = N \left[N - 2 \sum_{k=0}^{N-1} a_k \cos(\phi_k - \theta_k) + \sum_{k=0}^{N-1} a_k^2 \right]. \qquad (7.7)$$

The error is minimized if we choose $\phi_k = \theta_k$ for $k = 0, 1, \ldots, N - 1$. In other words, we choose H to have the same phase as F, while the magnitude of all coefficients of H are set to unity.

7.1.2.2 Signal Constellation

The procedure employed for generating the maximally separable sequences is as follows:

(1) From a random seed, generate a binary (± 1) sequence e_k of length $L = 2^{p_k - 1}$.
(2) Obtain the length-L_k DFT E_k of the binary sequence.
(3) Obtain S_k from E_k such that $|S_k(l)| = 1, l = 1, \ldots, L_k$ and $\angle S_k(l) = \angle E_k(l), l = 1, \ldots, L_k$.
(4) Take the length-L_k IDFT of S_k to obtain W_k. W_k is a *cyclic all-pass* function. All $L_k = 2^{p-1}$ cyclic shifts of W_k are orthogonal.
(5) W_k and the other $L_k - 1$ cyclic shifts of W_k, and their negatives, are the 2^{p_k} maximally separable sequences.

Note that the inner product of the sequence W_k of length L_k with each of the $2L_k = 2^{p_k}$ maximally separable sequences can be obtained by one length-L_k cyclic correlation efficiently implemented using the FFT. The index of the maximum absolute value of the cyclic correlation coefficients gives then detected sequence of p bits. Let $0 \le d_k \le 2^{p_k} - 1$ be the decimal representation of W_k^d.

$$W_k^d = \begin{cases} \alpha C(W_k, d_k) & \text{if } d_k < 2^{p-1} \\ -\alpha C(W_k, d_k - 2^{p-1}) & \text{if } d_k \ge 2^{p-1} \end{cases} \qquad (7.8)$$

where $C(x, q)$ stands for cyclic shift of the vector x by q (counter clockwise) positions, and α is a scaling factor that depends on Δ. For detection,

$$R_k = \mathcal{F}(\mathbf{W}_k)\mathcal{F}(\widehat{\mathbf{W}}_k) \quad r_k = \mathcal{F}^{-1}(R_k) \tag{7.9}$$

where $\mathcal{F}$ denotes the DFT, and

$$\tilde{d}_k = \begin{cases} \arg \max_{i=0\cdots L_k-1} |r_k(i)| & \text{if } r_k(i) > 0 \\ \arg \max_{i=0\cdots L_k-1} |r_k(i)| + L_k & \text{if } r_k(i) \leq 0. \end{cases}$$

An easier way of generating cyclic all-pass sequences $\mathbf{W}_k$ would be to generate them in the DFT domain by choosing unit magnitudes for DFT coefficients, but choosing the phases randomly. However, we need binary sequences of length $\frac{\Delta}{4}$ for the optimality of the practical type III schemes employed. Steps 1–4 ensure that the generated watermark signal (signature) $\mathbf{W}_k$ is an all-pass sequence *closest in the mean square sense* to the binary random sequence e_k.

The choice of the length L_k of each segment (which in turn decides the alphabet size) will depend mainly on the correlation ρ (the correlation measured between the embedded and extracted watermark signals) for the particular choice of embedding and detection parameters (i.e., Δ and β for thresholding type of postprocessing). Typically, the lower the value of ρ, the higher will be the value of L_k. Obviously, other factors like computational complexity may also influence the choice of L_k.

As the segment lengths are restricted to be powers of 2 for efficient implementation of the FFT, smooth trade-offs between bit rate and the probability of error can be achieved only by redundant signaling. In the next section, we propose a suitable and practical redundant signaling technique for improving the overall efficiency of the signaling method.

7.1.2.3   Redundant Signaling

For the proposed FFT-based signaling technique, we propose a combination of Reed-Solomon (RS) encoding [68] and introduction of parity for error correction. A sequence of d-bit symbols D_1 to D_n is RS encoded over $\mathcal{GF}(2^d)$, with block size of $2^d - 1$ (if $n < 2^d - 1$, the "shortened" code can be easily implemented by zero-padding $D_1 \cdots D_n$ to length $2^d - 1$ and considering the nonexistent symbols as "erasures" at the decoder). The RS-encoded sequence of d-bit symbols is then "appended" with q-parity bits to produce a p-bit symbol sequence, where $p = d + q$.

Signaling with parity can be done efficiently for the FFT-based technique. To introduce one parity bit (or reduce the valid points in the constellation by a factor of 2), we choose only odd values D between 0 and 2^{p-1} and only even values between 2^{p-1} and 2^p. This would correspond to choosing the largest from the *even-indexed* coefficients of r_k in Eq. (7.9). If $L_k = 2^{p-1}$ is the length of r_k, the even-indexed coefficients r_{e_k} of r_k can be obtained as (proof in Appendix C)

$$R_{2_k}(l) = R_k(l) + R_k(l + L_K/2), \quad l = 0, \ldots, \frac{L_k}{2} - 1$$

$$r_{e_k} = \mathcal{F}^{-1}_{L_k/2}(0.5 R_{2_k}). \tag{7.10}$$

In this equation, $\mathcal{F}^{-1}_{L_k/2}(\cdot)$ is an $L_k/2$-point IDFT (the factor 0.5 is irrelevant, as our intention is only to pick the coefficient with the highest magnitude). For introducing q-parity bits (in the segment L_k representing p bits, where $p = q + d$), valid points in the constellation are given by

$$D = \begin{cases} m2^q - 1 & D < L_k - 1 \\ m2^q & L_k \le D < 2L_k \end{cases} \quad m = 0, 1, \ldots, \frac{L_k}{2^q}. \tag{7.11}$$

In this case, only coefficients of r_k with indices that are multiples of 2^q are needed. For $l = 0, \ldots, L_k/2^q - 1$,

$$R_{q_k}(l) = \sum_{i=0}^{2^q - 1} R_k\left(l + i\frac{L_k}{2^q}\right) \qquad r_{q_k} = \mathcal{F}^{-1}_{L_k/2^q}(R_{q_k}).$$

Signaling with parity is especially useful for very low SNR data hiding (if ρ is very small, which results in large p or L_k).

For example, let $c \in \Re^{8192}$. For a low-noise scenario we may use segment lengths of $L_k = 64$ for each $p = 7$ bit symbol ($L_k = 2^{p-1}$). In such a scenario, we may use, for example, a block of RS code (127,111) over $\mathcal{GF}(2^7 = 128)$, which can correct up to eight errors in each block of length 127 (number of source bits $= 1$ block $\times$ 111 symbols per block $\times$ 7 bits per symbol $= 777$). However, if the SNR is low, we use segment sizes of $L_k = 1024$ ($p = 11$). If we do not employ parity bits, we need to use an RS code, say (2047, 2045). The maximum block size possible is, however, $8192/1024 = 8$. We need a shortened code. We may start with a source of six 11-bit symbols (66 bits), zero-padded to length 2045, and then perform (2047,2045) RS encoding, which can correct one error out of the eight transmitted symbols. Obviously this is computationally expensive. An alternative is to use $L_k = 512$ and $p = 10$, and also have $q = 5$ parity bits. We may now start with fourteen 5-bit source symbols (70 bits) and zero-pad it to a length-29 symbol block. This is followed by a computationally simple RS encoding (31,29). The first sixteen 5-bit symbols obtained after RS encoding are then made into 10-bit symbols by introducing 5 parity bits (which is done efficiently in the FFT-based method). For detection, the parity bits are stripped first to obtain a 16-symbol sequence of 5-bit symbols. This may be zero-padded to length 31 and RS decoded.

For data hiding applications in which computational complexity of detection is not a serious limitation or channel noise is low (implying small p), signaling with parity would be suboptimal. However, if p is large and $q = 0$ (or $d = p$), then RS encoding/decoding may become prohibitively expensive.

7.2 Synchronization

In some data hiding applications like image, video, and audio watermarking, preserving the synchronization between the embedding and detection operations becomes crucial. In such contexts, synchronization refers to the accuracy of the detector's information on spatial and temporal coordinates of the watermark signal in the stego signal. When the actual coordinates of the embedded watermark signal are different from the ones supposed by the extractor, detection performance may degrade significantly even though the traces of the watermark signal are present in the stego signal. Therefore, removing the synchronization between embedder and detector becomes a more effective attack than, say, attempting to "erase" the watermark signal from the stego signal. Geometric transformations like rotation, scaling, and translation (RST), warping, and signal cropping are the most common forms of desynchronization attacks [69], [70], [71]. For successful extraction of the watermark signal, data hiding methods require tools and techniques for restoring the synchronization efficiently, e.g., [72], [73], [74].

In the following sections, a hiding technique based on type III methodology with a thresholding type of postprocessing is proposed for watermark recovery from stego signals consecutively subjected to cropping and resizing operations. These attacks pose a threat of poor watermark detection due to signal transformation and signal loss. Hence, the detector has to be synchronized with the distorted stego signal prior to watermark extraction.

In general, if a particular desynchronizing attack can be modeled as a transformation, watermark detection could depend either on embedding in a domain that is invariant to that transform or on the ability to estimate the applied transformation by the attacker and invert it before detection. One particular technique that enables estimation of such transformation in the face of many different types of desynchronization attacks is periodic embedding and estimation of the transformation through cyclic autocorrelation.

It is shown that cyclic autocorrelation peak pattern (periodicity features of the signal) can specifically be used for calculating the resampling factor and estimating the amount of cropped data (i.e., number of deleted samples in a vector, number of pixels of line in an image). Therefore, the resampled signal can be restored to its original size.

The information loss due to cropping is countervailed by multiple embedding and redundancy coding of the watermark signal. Although multiple embedding is not an ultimate remedy to cropping, the motivation is that all replicas cannot be completely distorted simultaneously due to the perceptual constraints. Figure 7-1 is a representation of signal cropping and resampling. Erasures in the stego signal require reinstatement of synchronization. Synchronization is achieved by designing watermark

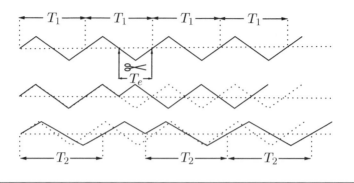

Figure 7-1   Representation of cropping and resampling consecutively.

signals in the form of all-pass filters that are orthogonal to all their cyclic shifts (Section 7.1.2.1). The phase of the all-pass filter is modulated by the message to be conveyed. RS error correcting codes are used for both introducing redundancy and achieving synchronization.

7.2.1   Autocorrelation for Restoring the Cropped Signal

Let a periodic signal $\mathbf{V}$ be obtained by combining n replicas of the signal $\mathbf{W}$ of length T_1 (Fig. 7-1). $\mathbf{V}$ is arbitrarily cropped out, $\mathbf{V}_C$, and the resulting signal is resampled by the factor $1/\tau = T_2/T_1$, $\mathbf{V}_{CR}$. Then, T_2 is the size of the resampled $\mathbf{W}$. Let n be a large integer number; T_e be the amount of signal (number of coefficients) cropped from $\mathbf{V}$, where $T_e < T_1$, and $L = nT_2 - T_e/\tau$ be the length of $\mathbf{V}_{CR}$. The resampling factor can also be defined as $1/\tau = L/(nT_1 - T_e)$. The autocorrelation $R_{V_{CR}V_{CR}}(m)$ of $\mathbf{V}_{CR}$ is computed as

$$R_{V_{CR}V_{CR}}(m) = \sum_{k=1}^{L-|m|} V_{CR}(k)V_{CR}(k+m). \tag{7.12}$$

In order to recover $\mathbf{W}$, the cropped resampled signal $\mathbf{V}_{CR}$ of size $nT_2 - T_e/\tau$ has to be restored to the cropped signal $\mathbf{V}_C$ with size $nT_1 - T_e$ by resampling with the factor τ. The autocorrelation function of $\mathbf{V}_{CR}$ is used to estimate $1/\tau$ depending on information about $\mathbf{V}$ available to the extractor (i.e., size of **V**, size of **W**). It will also be seen that the autocorrelation peak pattern provides insights into the nature of the croppings even when croppings occur at multiple positions (note that if two or more consecutive samples in **V** are cropped, they will be considered a single cropping). The total amount of cropped signal is assumed to be much smaller than the size of **V**, $T_e \ll nT_1$. The justification for this assumption is that in a typical attack scenario, due to perceptual constraints, the attacker cannot make radical changes on signal size **V**. Therefore, all copies of **W** cannot be cropped fatally at the same time. Consequently, in the corresponding autocorrelation function of $\mathbf{V}_{CR}$ the peaks observed at T_2 shifts of the origin, $R_{V_{CR}V_{CR}}(\pm iT_2)$, where $i \in \mathcal{Z}$, will be relatively greater in strength compared with other peaks, irrespective of the number of croppings. Given that T_1 is known at the extractor, the resampling factor can be found by measuring T_2 through distances between the dominant peaks in the autocorrelation function and calculating T_2/T_1. Alternately, if the size of **V** prior to cropping, nT_1, is known rather than the size of **W**, $1/\tau$ can be calculated using the relative peak locations of the autocorrelation function.

Considering the single cropping case of amount T_e, the autocorrelation function of the signal $\mathbf{V}_{CR}$ will indicate the presence of two periodic components with the same period, $T_2 = T_1 1/\tau$. The first component is identified by peaks at T_2 shifts of the origin. The second, on the other hand, generates peaks at the shift of $T_2 - T_e 1/\tau$ with respect to zero shift and at T_2 shifts thereafter. In other words, the first component is due to resampled copies of signal **W** in $\mathbf{V}_{CR}$, and the second one is due to the cropping. In the autocorrelation, at every $T_2 - T_e 1/\tau$ shift following a T_2 shift, the incomplete signal period coincides with a copy of itself and generates a peak. The peaks corresponding to the latter component are weaker in signal strength compared with the former due to the incomplete **W**. Therefore, other than the peak at the zero shift, every peak at T_2 shifts (with respect to zero shift) is accompanied by a peak due to cropped **W** (assuming n is large enough). The distance d between the peak at kT_2, $k \le n$, and $(k-1)T_2 + T_2 - T_e 1/\tau$ is calculated as

$$d = kT_2 - \left((k-1)T_2 + T_2 - T_e \frac{1}{\tau} \right),$$

$$= T_e \frac{1}{\tau}. \tag{7.13}$$

Being able to measure T_e/τ and T_2, the resampling factor τ is calculated as $\tau = nT_2/nT_1$ or $\tau = T_2/T_1$ based on availability of nT_1 or T_1. Then the total cropping amount T_e is calculated using Eq. (7.13). It should also be noted that given either of nT_1 or T_1, one can determine either using τ and T_2.

Now we shall consider the double cropping case where T_{e1} and T_{e2} are the amounts of the nonoverlapping cropped samples (T_{e1} and T_{e2} refer to croppings of $\mathbf{W}$ at different locations) from $\mathbf{V}$ with $T_{e1} + T_{e2} < T_1$. The autocorrelation function of $\mathbf{V}_{CR}$ may have up to four peaks in every T_2 interval that are $(k-1)T_2$, $k \leq n$, away from zero shift. These peaks may appear at $kT_2 - (T_{e1} + T_{e2})/\tau$, $kT_2 - (T_{e1})/\tau$, $kT_2 - T_{e2}/\tau$, and kT_2. The last one is due to resampled copies of $\mathbf{W}$ and has the highest correlation value. Others are due to cropped-resampled copies of $\mathbf{W}$ and have smaller strengths. If no croppings are present in the first and last periods of $\mathbf{W}$, for relatively large n and T_1, the distance, d, between the first and the last peak in any T_2 interval is measured as $(T_{e2} + T_{e1})/\tau$. Similar to the single cropping case, nT_2 and $1/\tau = nT_2/nT_1$ are consequently computed.

For more croppings followed by resampling, a similar analogy is applicable. If $T_{e1}, \ldots, T_{em}$ are the amounts of the nonoverlapping cropped signals and $T_{e1} + \cdots + T_{em} < T_1$, there may, at most, be $2m$ peaks at every shift based on how the signal $\mathbf{V}$ is cropped (i.e., the number of croppings in each period of $\mathbf{W}$, the location of a cropping in the period $\mathbf{W}$, the neighborhood of the cropped periods). These croppings may yield correlation peaks at 2^m locations in a T_2 shift (assuming each cropping is nonoverlapping with the others and considering that the first and last periods are not cropped). Corresponding peak locations in the autocorrelation function are at $kT_2 - \sum_{j=1}^{j=m} T_{ej}/\tau$, $kT_2 - \sum_{j=1, j \neq i}^{j=m} T_{ej}/\tau$ for $\forall i$, $kT_2 - \sum_{j=1, j \neq i,l}^{j=m} T_{ej}/\tau$ for $\forall i, l$ such that $i \neq l, \ldots, kT_2 - T_{ej}/\tau$ for $\forall j$, and at kT_2. Then, the distance d between the first and last peaks in a T_2 shift can be used to estimate the total erasure amount.

When the first and last periods of the signal $\mathbf{V}$ are cropped, the autocorrelation function may not generate a peak at $kT_2 - (T_{e1} + \cdots + T_{em})/\tau$. Therefore, the distance d, measured between the first and the last peak at a T_2 shift of the autocorrelation function, does not indicate T_e/τ. However, as will be explained in Section 7.2.3, d may still be measured using cyclic autocorrelation features for such croppings. Further, if both T_1 and nT_1 are known at the extractor, the amount of cropping, T_e, can also be determined by measuring d and $1/\tau$ using Eq. (7.13).

7.2.2 Practical Concerns

Calculating the resampling factor $1/\tau$ correctly depends on identifying correlation peaks and determining their relative locations in the autocorrelation function. However, some peaks may be buried in the correlation noise, which makes peak detection unreliable. Designing white noise like $\mathbf{W}$ signals and using cyclic autocorrelation are two remedies available for measuring d reliably.

7.2.2.1   Watermark Signal Design

The design of the signal $\mathbf{W}$ is critical, as autocorrelation properties of $\mathbf{W}$ characterize those of $\mathbf{V}$. Designing $\mathbf{W}$ as an all-pass filter which is orthogonal to all its cyclic shifts [67] gives one freedom to hide information by modulating the phase of the $\mathbf{W}$ as well as the improved autocorrelation properties (Section 7.1.2.1). An all-pass filter $\mathbf{W}$ of size T_1 gives $(T_1 - 1)/2$ degrees of freedom in modulating its phase, if T_1 is odd ($(T_1 - 2)/2$ degrees of freedom if T_1 is even).

7.2.2.2   Cyclic Autocorrelation

Cyclic autocorrelation enhances the correlation peaks due to signal wrapping in the autocorrelation function. Assuming $\mathbf{V}_{CR}$ has undergone multiple croppings of $T_{e1}, \ldots, T_{em}$, the corresponding cyclic autocorrelation can be obtained from the autocorrelation function by flipping the signal range $((nT_2)/2 - \sum_{j=1}^{j=m} T_{ej}/2\tau, nT_2 - \sum_{j=1}^{j=m} T_{ej}/\tau]$ and adding it onto signal range $(0, (nT_2)/2 - \sum_{j=1}^{j=m} T_{ej}/2\tau]$. After signal wrapping, the new coordinates for autocorrelation peaks in the range $((nT_2)/2 - \sum_{j=1}^{j=m} T_{ej}/2\tau, nT_2 - \sum_{j=1}^{j=m} T_{ej}/\tau]$ are found by subtracting their coordinates from $nT_2 - \sum_{j=1}^{j=m} T_{ej}/\tau$, which always coincide with one of the 2^m peak locations. For instance, if $\mathbf{V}_{CR}$ has been cropped once by removing T_e samples, autocorrelation peaks at kT_2 and $kT_2 - T_e/\tau$ for $k > n/2$ translate to $(n-k)T_2 - T_e/\tau$ and $(n-k)T_2$ in the cyclic autocorrelation function. For the general case, the peaks at $kT_2 - \sum_{j=1}^{j=m} T_{ej}/\tau$, $kT_2 - T_{ei}/\tau$ for $i \leq m$ and

(a)

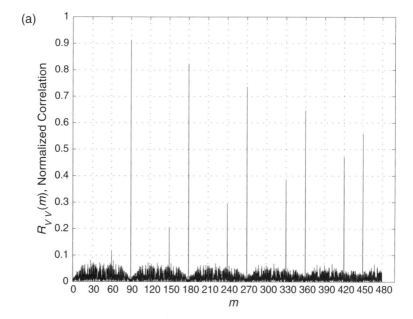

(b)

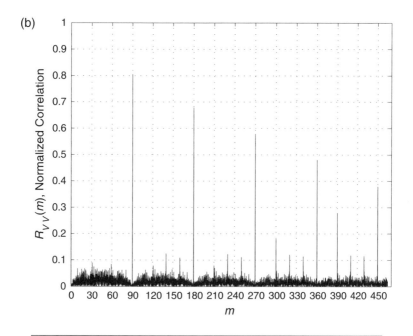

Figure 7-2 Computing total cropped amounts using cyclic autocorrelation $R_{V_C V_C}$. **(a)** Cropping once, $T_e = 20$. **(b)** Multiple cropping, $T_{e1} = 40$ and $T_{e2} = 20$.

kT_2, respectively, translate to $(n - k)T_2$, $(n - k)T_2 - \sum_{j=1, j \neq i}^{j=m} T_{ej}/\tau$ and $(n - k)T_2 - \sum_{j=1}^{j=m} T_{ej}/\tau$, making peak detection easier. Correspondingly, the autocorrelation peaks with the highest strength (due to cropped and resampled $\mathbf{W}$) will be translated to $(n - k)T_2$ and $(n - k)T_2 - \sum_{i=1}^{m} T_{ei}/\tau$ irrespective of the cropping pattern. Then, the resampling factor $\tau = T_1/T_2$ (or nT_1/nT_2) is reliably calculated by measuring the distance d between the two peaks, $T_e/\tau = (T_{e1} + \cdots + T_{em})/\tau$.

Figure 7-2a–b displays the cyclic autocorrelation functions, $R_{V_C V_C}(m)$, for single and double cropping cases. Signal $\mathbf{W}$ has a size of 90, and $\mathbf{V}$ is generated from 11 replicas of $\mathbf{W}$. In Fig. 7-2a, $\mathbf{V}_C$ is generated by cropping $\mathbf{V}$ once by removing the first 30 samples of the sixth period. On the other hand, in Fig. 7-2b, $\mathbf{V}$ is cropped twice by removing the middle 40 samples of the third period and the last 20 samples of the fifth period. In both figures, the peaks at multiple shifts of 90 (the size of $\mathbf{W}$) are easily identified, $\tau = 1$. Every shift of size 90, corresponding to the size of $\mathbf{W}$, contains two peaks in Fig. 7-2a and four peaks in Fig. 7-2b. The distance $d = T_e$, the number of erased samples, between the peaks in the former is 30 and between the first and fourth in the latter is 60.

7.2.3 Synchronization

The restored cropped signal must be repartitioned to recover $\mathbf{W}$. Since it is not certain which partitions are affected by cropping, the extractor needs some markers for reestablishing the synchronization. Most of the partitions contain signal $\mathbf{W}$ or a translated version of it. While some other partitions have cropped and translated versions of $\mathbf{W}$, RS error correcting encodes for generating $\mathbf{W}$ and handling synchronization. Since it is highly likely that most partitions will carry a cyclic-shifted version of $\mathbf{W}$, errorless decoding will be possible when the partition is reordered. Thus, given enough redundancy, both robustness to signal loss and synchronization are achieved, and errorless decoding of most of the partitions is possible at some cyclic shift of the partition.

7.2.4 Results

We implemented the methodology on a 512×512 graylevel Lena image, Fig. 7-3-a. Message m is assumed to be a sequence of 32 bits. The signal $\mathbf{W}$

takes the form of the watermark signal corresponding to m with a constraint on the correlation properties. Hadamard transform matrix is designated as the codebook and its orthogonal rows are mapped to codewords that are employed in watermark signal generation.

The message bit sequence is translated into words. Then the message words are redundancy coded using RS error correcting codes. Using the codebook, encoded message words are BPSK (binary phase shift keying) modulated and ordered in a way that fulfills the frequency domain symmetry requirements for the phase of the all-pass filter in order to generate the watermark signal $\mathbf{W}$. The watermark signal is chosen to be a 32×32 all-pass filter, which provides the hider with $\frac{32 \times 32 - 4}{2} = 510$ phase samples to modulate by the coded message m. Then, 16 copies of the watermark signal are embedded throughout the whole image.

The watermarked image is cropped, and in order to compensate the reduction in size, it is resampled back to its original size. At the extractor, a copy of the watermarked, cropped, and resampled image is divided into partitions of size $\mathbf{W}$. Watermark detection for each partition is followed by the two-dimensional cyclic autocorrelation of the detected set of signals. Using correlation peak pattern, resampling factor τ is estimated. The extractor, knowing an estimate of the total cropped amount but not their locations, resamples the image back to its size after cropping. Hence, the disturbing effects of the resampling can be reversed or at least minimized. This image is then repartitioned and the watermark extracted. Since extracted watermark signals may have been cropped and translated, an immediate detection of message m is not possible. RS codes are used to detect the message m from the extracted watermark signal, since they are capable of correcting burst error. The two-dimensional signal is shifted in rows and columns until an errorless decoding is possible. High redundancy coding helps in detecting message m even under severe signal loss.

Figure 7-3a–d displays the results for the described method applied on the Lena image, Fig. 7-3a. The watermarked Lena image is displayed in Fig. 7-3b, where the MSE per coefficient due to embedding is 6.9 (40 dB in PSNR). Figure 7-3c is the watermarked image cropped twice in both dimensions to a size of 488×488. Each cropping is the erasure of 12 lines of pixels in either horizontal or vertical dimension. The cropped image is

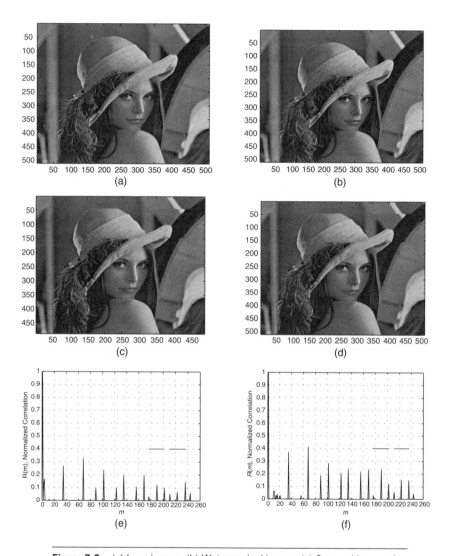

Figure 7-3 (a) Lena image. (b) Watermarked image. (c) Cropped image after watermarking. (d) Resampled image after cropping, and estimation of cropped amounts from the resampled image by projecting the two-dimensional auto-correlation function onto (e) horizontal dimension and (f) vertical dimension.

resampled back to its original size of 512×512 in 7-3d. Figure 7-3e–f shows the projections of the cyclic autocorrelation function onto horizontal and vertical dimensions. The distance between the first and last peaks in each period, corresponding to the size of the watermark signal enlarged by the resampling factor, of the cyclic autocorrelation function is $d = 25$, which has an estimation error of 1 line of pixels in both dimensions. T_2 is also measured using the Fig. 7-3e–f as 33 at some shifts and as 34 at most of the others, $\frac{32}{34} < \tau < \frac{32}{33}$. The image in Fig. 7-3d is resampled to a size shorter by 24 ($T_e = \text{round}(25 \times \frac{32}{34})$) lines of pixels in each dimension, partitioned in 32×32 blocks and watermark detected. Extracted signals from each block are averaged. Then the averaged signal block is decoded in cyclic shifts of rows and columns until an errorless decoding is possible. For the presented implementation, the redundancy rate is around $\frac{1}{15}$ $\left(\frac{32}{510}\right)$. RS codes were successful in detecting the 32-bit message m with no errors.

7.3 Perceptual Constraints

As the resource of the communication between the hider and the attacker is the total imperceptible distortion that can be introduced into a given host signal, achieving the optimal rate vs robustness performance requires a higher-level understanding of the host signal in the perceptual sense. Data hiding methods, most generally, approach the problem by incorporating simplified perceptual models or the findings of perceptual compression with the embedding process.

Most elaborate formulations of data hiding (as discussed in this chapter) rely on a fixed distortion measure, e.g., MSE distortion, for analytical tractability. Hence, the corresponding analyses and results oversimplify this aspect of the problem. Evaluated from an imperceptibility perspective, type I methods can exploit the host signal information better than type II or type III methods.

Within the additive schemes, embedding is done by adding a scaled version of the watermark signal to the host signal or to a transformed version of it. The proper weighting for each watermark signal sample can be locally determined according to just-noticeable-difference thresholds and masking principles, thereby complying with perceptual constraints. In quantization-based techniques, however, the distortion introduced into each host signal coefficient can be controlled only in an indirect manner, by adjusting either the quantization step size or the amount of processing distortion. Since the optimization procedure for the embedding and detection parameters assumes power-limited distortion, which disregards the perceptual properties of the host signal, the corresponding embedding operation is nonoptimal in terms of perceptual criteria. In this respect, scalar quantization-based embedding/detection schemes provide a better control, since each coefficient is embedded individually and Δ or the postprocessing parameter can be selected to comply with perceptual constraints, whereas in schemes that employ high-dimensional quantization, the introduced distortion due to embedding is minimized over the quantized vector, which would not necessarily limit the distortion introduced into each coefficient.

In order to achieve imperceptibility, type II and type III methods select the power constraint conservatively. This leads to an underutilization of the communication resource. Compared with type II methods, the postprocessing involved in type III methods gives the hider another degree-of-freedom into controlling the distortion introduced into each host signal sample. Hence, the embedding parameter that designates the amount of processing distortion introduced into the quantized host signal (i.e., β in thresholding, α in distortion compensation, σ_V in Gaussian mapping) can be fine-tuned in accordance with the perceptual features of the host signal. Thresholding and distortion-compensation types of postprocessing can be readily adapted to applications with more strict imperceptibility requirements through adjusting β and α, whereas with Gaussian mapping, modulating the processing distortion is a more complex task due to the nonlinear transformation. However, the optimal approach is to revise the optimization procedures given in Section 5.2 (Eqs. (5.33), (5.37), and (5.39)) by taking into account the perceptual properties of the host signal as constraints (rather than limiting the distortion power to P) during the optimization of embedding/detection parameter values. This is probably the next step for future research in quantization-based embedding methods.

7.4   Attacks on Data Hiding Systems

The vast majority of research in the field of information hiding is motivated by ownership protection, copy and access control, and authentication of digital media content, e.g., image, audio, video. Common to all of these motives is the likely presence of intruders willing to modify contents that have undergone processing, with the intention of nullifying aforementioned efforts. Correspondingly, the research focus in this direction is the design of information hiding methods that can survive very sophisticated attacks while exploiting the perceptual characteristics of the cover signal to the fullest. In this setting, robustness of a scheme, rather than its capability to convey maximum amount of information, is central to the evaluation process. This has drawn a lot attention to the design of optimal attack strategies. In this section, the conventional and the state-of-the-art attacks are reviewed and categorized by highlighting the research direction in this field.

Attacks on a stego signal intend to impair the detector's ability to extract the message signal through all possible means without perceptually modifying the stego signal. The research in attack design grows at a pace that parallels the improvements in embedding/detection techniques [69], [70], [75], [76], [77], [78], [79]. The initial phase of attacks evolved mainly by applying common signal processing operations. This type of operation lacks the ability to apply the optimal attack, as it does not fully exploit any prior information on the embedding process or cover signal but, rather, attacks in a blind manner. However, these attacks have led to great success by pointing out common deficiencies of many methods and instigating improvements to resist such manipulations.

From the game-theoretic point of view, as put forth in [42] and [43], the solution of the game between hider-extractor and attacker depends on the attacker's information on the hider's strategy. Therefore, designing attacks that do not utilize any available information on the embedding scheme, cover signal, and embedding distortion may yield overly optimistic results for the method. The type of attacks that intend to compensate for this drawback, by more effective design, constitute the second phase [77], [79]. These attacks rely on estimating either the original cover signal or the watermark signal from a given stego signal generated by an additive embedding scheme. Then the stego signal is attacked by either removing the watermark signal or remodulating the stego signal so that the attacked signal has no traces of the watermark signal.

The attacks on information hiding systems can be classified into four categories based on how the watermark extraction operation is impaired. These are:

- Removal attacks

    - Blind attacks
    - Estimation attacks

- Desynchronization attacks
- Security/cryptographic attacks
- Protocol attacks

In the following sections, each category is briefly described.

7.4.1 Removal Attacks

This type of attack aims at removing the watermark signal so that a conclusive extraction of the signal is not possible. Based on the underlying principle for removal, these attacks can be grouped into two kinds: blind attacks and estimation attacks. The former assumes that the stego signal is a distorted version of the cover signal and tries to remove the noise without any reference to the original, whereas the latter models the cover signal and the embedding distortion statistically and removes the noise by proper filtering.

7.4.1.1   Blind Attacks

(1) Noise Addition. A random noise is added to the stego signal in order to garble the embedded watermark signal. The success of the attack depends on the power and the correlation of the noise with the embedding distortion.

(2) Digital(D)/Analog(A) to A/D Conversion and Quantization. These are effective against embedding schemes that do not effectively exploit the cover signal's features for embedding; e.g., embedding by manipulating LSBs of the cover signal, even/odd modulation.

(3) Filtering. A simple filtering operation can be used to remove the signal content in the perceptually insensitive bands. This attack can be effectively used to distort the pseudo-noise-like and high-pass watermark signals.

(4) Statistical Averaging and Collusion Attacks. When the attacker has access to multiple copies of a cover signal watermarked with different keys or watermark signals, the attacker can successfully produce a signal with no watermark signal. If the embedding distortion is independent of the cover signal, averaging the stego signals will cause the watermark signals to cancel each other or will generate an unrecognizable noise signal. Alternately, the attacked signal can be generated by combining different parts of the stego signal.

(5) Perceptual Quantization. A quantization tuned to perceptual features of the cover signal may remove all the redundancy in the stego signal. Thus, the watermark signal can be removed or impaired severely.

(6) Multimedia Processing. Common multimedia processing techniques used in analyzing media data provide powerful means to modify the contents. For example, image processing operations like blurring, sharpening, edge enhancement, despeckling, histogram modification, Gamma correction, brightness and/or contrast changes, and color quantization can be applied on the stego signal in order to render the watermark signal undetectable. Similarly, the watermark signal embedded in a stego audio signal can be distorted by processings like time stretching, zero cross insertions, copying/cutting samples, and reverbing.

7.4.1.2 Estimation Attacks

In estimation-based attacks, the attacker assumes priors on the original cover signal and the watermark signal. Then, using one of the stochastic criteria such as MAP (maximum a posteriori), ML (maximum likelihood), or MMSE (minimum mean squared error criterion), the original signal or the embedding distortion is estimated.

(1) Denoising Attack. The cover signal is estimated and the stego signal is replaced by the obtained estimate.
(2) Remodulation Attack. The embedding distortion is estimated, and then the modulation applied to the cover signal in generating the stego signal is reversed.

7.4.2   Desynchronization Attacks

The attacks in this class intend to remove synchronization between the embedder and the detector rather than removing the watermark signal. Therefore, the detector is, ideally, able to recover the embedded signal once the desynchronization attack is identified and reversed, although this may be a very high complexity task.

7.4.2.1   Geometric Attacks

These attacks include common image processing manipulations applied on a global and local scale. Some of the most common manipulations are RST change in aspect ratio, shearing, line or column removal, cropping, random alterations (e.g., random bending), and jittering. Unzign and StirMark are two popular software packages that enable applying a combination of these attacks to a given stego signal.

7.4.2.2   Mosaic Attack

This attack relies on the assumption that an extractor will not be able to extract the watermark signal from a randomly chopped piece of a stego signal. The attacker utilizes this in web-based applications by displaying a stego signal as a tile of many pieces, without hindering the visible quality, in order to trick the webcrawlers.

7.4.2.3   Template and Periodicity Removal

If the synchronization between the embedder and the detector is established based on a template or a periodicity feature of the stego signal, the attacker can simply detect and erase them. Hence, the extractor will be vulnerable to simple geometric attacks.

7.4.3 Security/Cryptographic Attacks

Unlike the previous two categories, this class of attacks do not intend to remove the watermark signal or fail the extractor's operation in a direct manner. Rather, they try to find a secret of the employed method by exhaustive searches for manipulating the stego signal or the extractor, e.g., brute search of the secret key.

> (1) Oracle Attack. When the watermark extractor is available, the attacker can remove the watermark signal by introducing small changes to the stego signal until watermark extraction fails. With this attack, the attacker ensures that the stego signal is distorted no more than required to remove the stego signal.
>
> (2) Software and Hardware Attacks: An attacker can decompile an available software or reverse-engineer tampered hardware in order to obtain some valuable information or to disable certain functionalities.

7.4.4 Protocol Attacks

Protocol attacks aim at shedding doubt on the reliability of an information hiding system.

7.4.4.1 Invertible Watermarks

If a stego signal is generated by an invertible watermark, the attacker can create ambiguity in the ownership by subtracting his watermark signal from the stego signal and claiming the resultant signal as his original.

7.4.4.2 Copy Attack

The attacker can use a form of estimation attack to estimate the watermark signal embedded in a stego signal and then transfer it to another signal, thereby creating confusion.

7.4.5 Future Direction in Attack Design

It should be noted that an attacker will usually apply a combination of the previously mentioned attacks, thereby exposing the stego signal to a more vulnerable situation. This requires more sophisticated information hiding system designs with a better utilization of the perceptual characteristics of the cover signal, which will induce a similar action in the attack design process. An immediate further step in this direction is to adopt more sensitive distortion measures rather than the widely used MSE distortion, e.g., Watson's metric. Another issue is the extension of estimation-based attacks to more general embedding schemes, other than linear additive schemes. Finally, the theoretical aspects of the attack design need to be addressed within a game-theoretic framework for the practical data hiding applications. This will have a deep impact on the study of information hiding.

Data Hiding Applications

In this chapter we explore some applications of data hiding. Even though there are numerous potential applications of data hiding, our explorations are restricted to three common applications that need robust data hiding. It is pertinent to mention here that there is a class of applications of data hiding in which the hidden data is rendered very fragile on purpose. Obviously, such applications are not based on the theoretical framework presented in the earlier chapters, which try to establish techniques to maximize the capacity or robustness of data hiding.

In particular we address three applications of data hiding. In Sections 8.1 and 8.2 we consider applications in which only robustness to compression is needed. In Section 8.3 we consider applications in which data hiding is required to be robust to many generic attacks. Specifically, we address the issue of using the hidden data to resolve ownership of watermarked content.

8.1 Design of Data Hiding Methods Robust to Lossy Compression

In this section, we begin by exploring the intricacies of the duality of data hiding and data compression to help develop efficient data hiding techniques for images that can reasonably resist lossy compression. The problem of efficient data hiding is divided into two subproblems. The first is maximization of the *resource*, which is the *permitted distortion* of images. The second is the *efficient use of the resource* by means of sophisticated *signaling techniques*. Various solutions for both problems are proposed, and their advantages and disadvantages are discussed. We conclude that a good solution to the first problem is to choose *magnitude DFT* as the domain in which the message signal is embedded. For the signaling method, we use a type III scheme, which uses periodic functions to tile the space of magnitude DFTs of images with a large dimensional signal constellation. The large dimensional signal constellation is in turn efficiently implemented using the FFT.

Applications of data hiding can be classified in many ways. One classification may be based on the robustness requirements of the data hiding application. For instance, applications like watermarking typically require robustness to *intentional tampering*. On the other hand, some applications may need robustness to only *unintentional* attacks (attacks not especially directed at removing the hidden data), like lossy compression. Another classification may depend on the *restrictions* to be placed on data hiding. For example, *invisible watermarking* is expected to resolve rightful ownership of the multimedia content, unambiguously, in a court of law. For this purpose many restrictions may have to be imposed [80], [81] [82], [83] on data hiding for watermarking. In contrast, virtually no restrictions are placed on applications like *secret communications* (communication between two private parties through a *subliminal channel* facilitated by data hiding). We focus on data hiding applications and methods for images and video. We also *restrict ourselves to applications that require robustness to only lossy compression.*

8.1.1   Data Hiding for Secure Multimedia Delivery

Data hiding is expected to be a boon for multimedia content providers, who can expect to communicate with *compliant multimedia players* through the subliminal channel provided by data hiding. This communication could control access and provide customized delivery and solutions for pay-per-view implementations [6], [84]. A compliant multimedia player

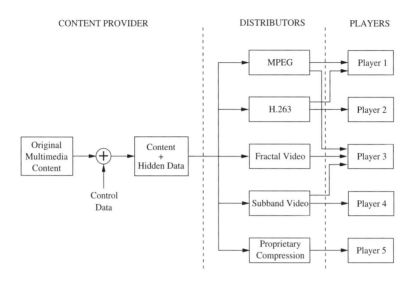

Figure 8-1   Block diagram of a multimedia distribution system. Though the generic multimedia players may support only a limited number of compression formats, all the players follow the same protocol for extracting the hidden control information. Player 3 supports three different formats, while player 5 supports only the proprietary compression format.

would honor an agreed-upon protocol for extracting (and abiding by) the hidden control information. Figure 8-1 displays a block diagram of a possible multimedia delivery system. Content providers (the creators of multimedia content) can hide pertinent control information for the multimedia players and make it available for distribution. The distributors may compress the content using some standard or proprietary compression method before it reaches the end users (or their multimedia players). The content may be distributed by several distributors in different formats, understandable by different players. However, as long as all such players follow an established protocol for extracting the hidden information, *and the hidden data is able to survive all the lossy compression methods employed by the distributors*, the content providers can indirectly *control* compliant *players through the hidden information*. Hiding the information in the raw multimedia data ensures that the hidden data stays embedded forever in the content.

Unless the hidden data is extracted with a "reasonable degree of certainty," the compliant multimedia players may refuse to play the content. Thus intentional tampering for the purpose of removing the hidden information serves to make only that *particular copy* of the content unusable. On the other hand, the motivation to make it robust to *all* compression methods is to facilitate more efficient distribution of the content. Failure of the hidden data to survive a "good" compression method makes that compression method unusable for distributing that content.

8.1.2 Compression and Data Hiding

Multimedia compression tries to convey the information in a multimedia content as efficiently as possible, with the fewest number of bits. Data hiding, on the other hand, tries to sneak additional bits of information into the original content. As the "additional information" does nothing to improve the quality of the content, an ideal compressor would completely suppress the hidden information.

Let $\mathcal{I}$ represent the space of $M \times N$ images of b bits per pixel (2^{MNb} possible images). Alternately, every point in $\mathcal{I}$ is an $M \times N$ image. As the image is represented by fewer bits in the compressed domain, many original image points are mapped by the compressor to one image point after (lossy) compression (and decompression). As an example, in Fig. 8-2, all points in the range R are mapped to a single point D.

In the figure, consider an image A (represented by +) in the region R. Let us assume that we want to hide one bit of information in the image A that would survive compression. The space $\mathcal{I}$ is *completely* tiled by two regions that represent 0 or 1. For example, if the image A is located in a region representing 0, it could be left intact if the bit to be hidden is 0. To hide a bit 1, however, A has to be moved to a point B (represented by ∗), which simultaneously belongs to region 1 and lies *outside the range R*, so that after compression (and decompression) the image is mapped to a different point B_1. For hiding n_b bits in an image that can survive compression, the image has to be distorted such that after decompression the image is mapped to any of 2^{n_b} possible points. In this case, the space of images has to be tiled by 2^{n_b} regions.

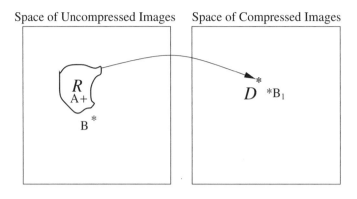

Figure 8-2 A lossy compression/decompression sequence maps all points in the range *R* to a single point in the domain *D*.

Now, it is easy to see that no data hiding would be possible with an *ideal* compressor. If δ_t is the visual distortion permitted (δ_t *may not* be a measure of the MSE), then there exists a finite number of points to which the original image may be "moved." However, an ideal compressor with the same threshold $\delta_c = \delta_t$ would map all such points to a *single point* in the space of decompressed images! Unless we employ different standards (a measure of δ) for the quality of the image after data hiding and for the decompressed image (or unless $\delta_c > \delta_t$) *no data hiding would be possible with ideal compressors.* However, practical compression techniques are not ideal. Therefore, efficient design of data hiding should utilize the *holes in compression techniques.*

8.1.2.1 Data Hiding with Known Compression

When the compression method the image is likely to undergo is known in advance, it is easier to design efficient data hiding methods. For example, let us assume that it is known in advance that the images will undergo only DCT-based JPEG compression with the *default quantization matrix.* Let us also assume that the image is not expected to undergo compression more severe than quality factor 50%. The best data hiding method for such a situation would be the following [85]:

- Let $Q(m, n), m = 1, \ldots, 8, n = 1, \ldots, 8$ be the quantization matrix for JPEG at 50% quality. The matrix is tabulated in Table 8-1.

TABLE 8-1
The DCT Quantization Matrix $\mathbb{Q}$

16	11	10	16	24	40	51	61
12	12	14	19	26	58	60	55
14	13	16	24	40	57	69	56
14	17	22	29	51	87	80	62
18	22	37	56	68	109	103	77
24	35	55	64	81	104	113	92
49	64	78	87	103	121	120	101
72	92	95	98	112	100	103	99

- Let K be the total number of coefficients (among the $M \times N$ 2-D DCT coefficients of the image) that quantize to a *nonzero* value when the quantization matrix $\mathbb{Q}$ is used.
- Let b_s be a bit sequence of length K to be hidden in the image.

Fix a particular scan order for the $\frac{M}{8} \times \frac{N}{8}$ image blocks. Fix a scan order for the 8×8 coefficients of each block. We hide one bit in each nonzero (after quantization) coefficient (as a significant amount of compression is achieved by JPEG compression due to efficient run-length coding of the coefficients that quantize to zero, changing coefficients that quantize to zero would affect the compression ratio of the image with embedded data). Let $\mathbf{C}$ be the vector of nonzero coefficients. For $i = 1, \ldots, K$, if $b_s(i) = 0$, then force the coefficient $\mathbf{c}(i)$ to quantize to an odd number. Otherwise force it to quantize to an even number. If the values are forced to the midpoints of the quantizers, then the hidden data would survive JPEG compression of any quality as long as it is better than 50% (if they are not forced to the *midpoints* of the quantizer steps, the hidden data will survive JPEG-50 but may not survive any higher quality compression, like JPEG-75!). For extracting the hidden information, the DCT of the image blocks (of the received image) are obtained. The DCT coefficients are quantized using the quantization matrix $\mathbb{Q}$. All coefficients quantizing to zero are ignored. All other coefficients are arranged in the prescribed order. If the quantized result is odd, the hidden bit is a zero. Otherwise, the hidden bit is a 1. Figure 8-3 depicts the achievable data hiding capacities for 11 standard

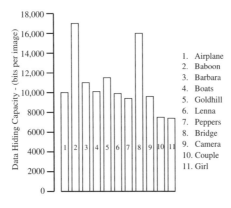

Figure 8-3   Data hiding capacities (number of DCT coefficients that quantize to a nonzero value with quantization matrix $\mathbb{Q}$) of 11 256 × 256 test images.

test images using this simple data hiding technique. However, the hidden data is very unlikely to survive other forms of lossy compression, even if DCT-based JPEG is used with a different quantization matrix.

8.1.2.2   Simultaneous Robustness to Multiple Compression Techniques

Consider the space $\mathcal{I}$ of original images. When the compression method is known (as in the previous section), we make use of the fact that points (or "states") R_1 to R_n are mapped to the same points R_1 to R_n in the space of decompressed images, as shown in Fig. 8-4. Therefore, the number of valid states of the compression method that lie within an envelope of "unnoticeable visual distortion" is a direct measure of the number of bits that can be hidden in an image (in the previous example, it is the number of valid JPEG-50 compressed images within the envelope of unnoticeable visual distortion).

The problem becomes more complicated if the hidden data has to survive *multiple* compression methods. Consider three compression schemes $\mathcal{C}_1$, $\mathcal{C}_2$, and $\mathcal{C}_3$. In Fig. 8-5 + denotes points in $\mathcal{I}$ which are permissible $\mathcal{C}_1$-compressed (and decompressed) images. Similarly, • and ∗ stand for $\mathcal{C}_2$- and $\mathcal{C}_3$-compressed images, respectively. Let A be the original image and $\boldsymbol{R}$ be an envelope of the possible points that A could be moved to with

SPACE OF ORIGINAL IMAGES   SPACE OF COMPRESSED IMAGES

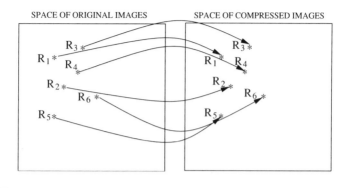

Figure 8-4   Known compression method.

R

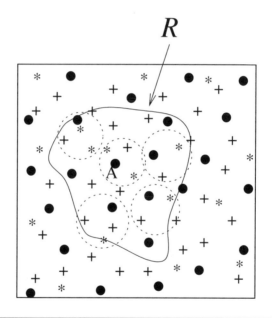

Figure 8-5   Data hiding with robustness to different compression methods.

unnoticeable visual distortion. If the hidden data has to survive *any* compression scheme, then the number of possible states (2^p, where p is the number of bits that can be hidden) is limited to the number of nonintersecting regions (marked by dotted circles) where at least one of the valid points of each compression scheme can be found.

8.1.2.3   Robustness to Unknown Compression Methods

However, if the exact effect of compression is not known (the valid states are not known *a priori*), the job of designing efficient data hiding methods warrants a totally different approach. Since one has no idea of the "valid" compression points (or valid compressed images for that particular compression method), the centers of the nonintersecting regions have to be considerably well separated to ensure that at least one valid compression point of all compression methods lies in each hypersphere. However, the following questions arise:

- A large distance between the centers of the hyperspheres implies that it may be necessary to introduce a significant amount of distortion to move the image to a desired state. Is it possible to do that without affecting the visual fidelity of the image?
- Assuming that it is possible to introduce a significant amount of distortion without affecting the visual fidelity to move the image A to a new point $\hat{A}$, why should a good compressor map two *visually identical* images A and $\hat{A}$ to different points in the compressed domain?

The answer to the second question is the following. All known compression methods try to minimize the MSE between the original and the compressed image—in the new generation of compression methods (like EZW, SPIHT, and IFS [fractal] image compression) even more so than the DCT-based JPEG. This is a hole *common to all compression methods* and can be used effectively for data hiding if satisfactory answers to the first question exist. In the next section we explore solutions to the first question.

8.1.3   Utilizing the "Hole" in Compression Techniques

As stated in the previous section, if the images can be modified considerably in the mean square sense without affecting the visual fidelity of the image, then one can achieve large separation between states corresponding to different bit sequences, and thus achieve robust data hiding.

One solution to this problem (of trying to introduce as much distortion as possible without affecting the visual fidelity) is to use good models of visual thresholds (for example, see [86], [87]) to embed the

Original Image   After Histogram Modification   After StirMark

Figure 8-6 Left: Original Goldhill image. Center: Goldhill image obtained by modifying the histogram. Though both images look similar and are of good visual quality, the difference between the two images in terms of PSNR is 20 dB. Right: Image obtained after StirMark. The difference between the two images in terms of PSNR is 19 dB.

hidden bits. Many data hiding methods [88] that utilize these models have been proposed. However, the main drawback of these methods is that well-defined visual threshold models (e.g., in the DCT or wavelet domain) also suggest the compression-technique means to improve their performance. Thus, when one uses these models to add a significant amount of signature energy to certain coefficients of the image, a better compression technique that may evolve in the future may also make use of these visual thresholds to perhaps *quantize those coefficients more coarsely.* In other words, *utilizing such visual threshold models indirectly amounts to utilizing holes that can be easily "plugged" in the future.* One of the main advantages of data hiding is that the hidden data stays with the content *forever.* As compression techniques improve in the future, content *distribution* becomes more efficient. But if the hidden data is not able to survive those compression methods, the content loses its value. Therefore, more useful data hiding techniques should utilize holes that are very difficult to plug.

Figure 8-6 depicts the original 256×256 Goldhill image, its histogram-reshaped version, and the image after StirMark [71] (StirMark is a watermark attack software that introduces imperceptible geometric distortions in the image). Though the second and third images are very close to the original in visual fidelity, their PSNRs are 20 and 19 dB, respectively! It is clear that significant amounts of distortion (in the MSE sense) can be tolerated as long as the introduced distortion *modifies only the histogram or introduces small geometric distortions, or perhaps both*. Hence, if we are able to embed the hidden data by introducing geometric distortions and/or histogram modification, a large separation between different states can be obtained.

However, things may not be as simple as they seem at first glance. Let $\mathbf{I}$ and $\mathbf{I}_1$ be two "similar" images (assume that $\mathbf{I}_1$ is derived by lossy compression of $\mathbf{I}$). Let $\mathcal{H}(\mathbf{I})$ be a function of the histogram of the pixels of image $\mathbf{I}$. If we try to embed data by *specifying* $\mathcal{H}(\mathbf{I})$ [89], the hidden data will *not* be robust to compression. Even small modifications in the MSE (like what may typically be introduced by lossy compression) can change the histogram significantly. Similarly, if $\mathcal{G}(\mathbf{I})$ is a function of some geometric features[1] of the image $\mathbf{I}$, and d is some metric, $d(\mathcal{G}(\mathbf{I}), \mathcal{G}(\mathbf{I}_1))$ may be large even if $d(\mathbf{I}, \mathbf{I}_1)$ is small. Just as introduction of small geometric distortions can cause a significant change in the MSE, introduction of small distortions in the MSE may cause significant changes into $\mathcal{G}(\cdot)$. This is the reason that the watermarking technique proposed by Rongen *et al.* [90] is robust to StirMark, but not very robust to JPEG compression. In order to achieve robustness to compression, the well-separated states (corresponding to the bit sequence to be embedded) have to be *specified first*. Then geometric distortions and/or histogram modifications have to be introduced to move the image close to the specified state. However, there may not be a simple or even methodical way to do this. But if such a method can be found and implemented with a reasonable degree of computational complexity,[2] it promises to be an excellent solution to the problem of robust data hiding.

A practical way to introduce a large amount of distortion into the image without affecting its visual fidelity is to modify only the magnitudes of the DFT coefficients. Figure 8-7 (left) shows the original 256×256 Boats image. The center image (14.1 dB PSNR) was derived by retaining the DFT phases of the original image and choosing random magnitudes. In spite of the very low PSNR of the image, we see that a significant amount of "information" about the original image is preserved. The third

---

[1]For example, in [90] the function depends on the spatial location of "salient" features.

[2]Computational complexity of the data embedding algorithm is not a serious limitation for the applications proposed in Section 5. Data embedding is done only once for each content.

Figure 8-7   Left: Original Boats image. Center: Boats image obtained by retaining the DFT phases of the original image and choosing random magnitudes (PSNR 14.1 dB). Right: Image obtained by retaining DFT magnitudes of the original and choosing random DFT phases (PSNR 15.6 dB).

image (right, 15.6 dB PSNR) was derived by retaining the magnitudes of the DFT coefficients of the original image but choosing the DFT phases randomly. Even though the PSNR of the third image is 1.5 dB better than that of the image in the center, the resulting image conveys almost no information about the original. This illustrates the well-known fact that the human visual system is much more sensitive to DFT phase than DFT magnitudes [91].

Thus if the data embedding is done in the magnitude-DFT domain (the states are specified by their magnitude-DFT coefficients—embedding the data changes the magnitudes of the DFT coefficients of the original image but leaves the phase intact), a significant amount of distortion (in the MSE sense) can be introduced without affecting the visual fidelity of the image. In addition, unlike the use of well-defined visual threshold models, this is not a hole that is capable of being easily plugged in the future (compression techniques that utilize the DFT and quantize the magnitudes coarsely and the phases finely have been proposed but have not been effective [92], [93]).

Introducing the distortion into the magnitude-DFT coefficients (for embedding information bits) can be achieved as follows. Let $\mathbf{I}$ be the original $M \times N$ image. Let $\mathbf{I} \stackrel{\mathcal{F}}{\longleftrightarrow} \mathbf{I}_{\mathcal{F}}$, where $\stackrel{\mathcal{F}}{\longleftrightarrow}$ stands for 2-D DFT pairs. $\mathbf{I}_{\mathcal{F}}$ has 4 real coefficients and $MN - 4$ complex coefficients. Only half ($D_0 = \frac{MN-4}{2}$) of them, however, have unique magnitudes. Let $\mathbf{C}_{I_{\mathcal{F}}} \in \Re^{D_0}$ be a vector of the unique magnitudes of the complex DFT coefficients of $\mathbf{I}_{\mathcal{F}}$. Every image can be represented as a point in D_0-dimensional space.

The D_0 magnitude-DFT coefficients serve as the carriers for the subliminal communication. However, as high-frequency DFT coefficients may not be able to survive lossy compression, we shall use only a subset $\mathbf{C} \in \mathfrak{R}^D$ of $\mathbf{C}_{I_\mathcal{F}}$ for data hiding.

8.1.4 The Data Hiding Scheme

Figures 8-8 and 8-9 show the block diagrams of data embedding and the data detection schemes. The figures are self-explanatory, except for the additional "Key-Based Transform" blocks. A truly secure data hiding scheme should be difficult to crack even if every step of the algorithm for data hiding were public. In this case, the only "secret" should be the key $\mathcal{K}$ (even though it is possible to have Δ as part of the key, as its choice is demanded by design criteria, one would not have very much freedom in choosing Δ). If the transform employed (DFT) and the value of Δ are public, then the signature can be easily "read," especially if binary signatures are used. While erasing hidden data may not be a very serious issue for multimedia delivery, *modifying* it may have disastrous consequences. Security can be vastly improved by using a key-based transform before data embedding (and therefore before detecting). In the proposed scheme, we use a simple key-based transform based on cyclic all-pass filters.

Let $\mathbf{h} \overset{\mathcal{F}}{\longleftrightarrow} \mathbf{H}$, where $\mathbf{h} \in \mathfrak{R}^N$ is cyclic all-pass (or $|H(k)| = 1 \, \forall k$). As all cyclic shifts of $\mathbf{h}$ are orthogonal, they form a basis for $\mathfrak{R}^N$. The phases ϕ_n, $n = 0, 1, \ldots, N - 1$ of the elements of $\mathbf{H}$ can be arbitrary, and we have N degrees of freedom for choice of the vector $\mathbf{h}$ with mutually orthogonal circular shifts. For real $\mathbf{h}$ we have $\frac{N}{2} - 1$ phase values that can be arbitrarily chosen. Thus, a pseudo-random all-pass sequence of length N can be generated from a pseudo-random (uniformly distributed between π and $-\pi$) sequence of length $\frac{N}{2} - 1$. The pseudo-random sequence would be generated from the key $\mathcal{K}$. If

$$\phi_k = \begin{cases} 0 \text{ or } \pi & k = 0, k = \frac{N}{2} \\ \theta_k & k = 0, \ldots, \frac{N}{2} - 1 \\ -\theta_{N-k} & k = \frac{N}{2} + 1, \ldots, N - 1 \end{cases}$$

$$H(k) = \cos(\phi_k) + \mathrm{i} \sin(\phi_k), k = 0, \ldots, N - 1 \qquad (8.1)$$

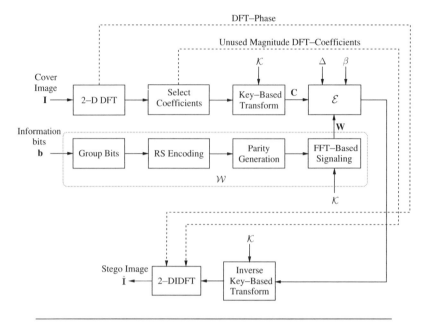

Figure 8-8 Block diagram of data embedding.

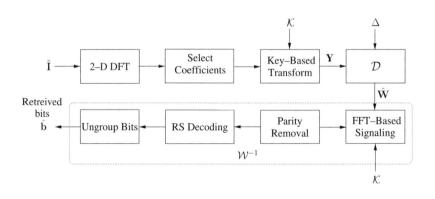

Figure 8-9 Block diagram of data detection.

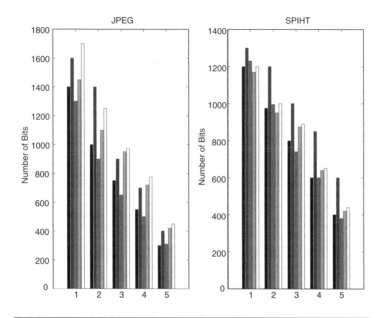

Figure 8-10 Plots of achieved data hiding capacities for JPEG (left) and SPIHT (right) compression for five 256 × 256 test images (Lena, Barbara, Boats, Goldhill, and Girl). JPEG compression scenarios 1–5 correspond to quality factors 75, 65, 55, 50, and 40, respectively. SPIHT compression scenarios 1–5 correspond to 1.35, 1.25, 1.15, 1.10, and 1.0 bpp, respectively.

where $i = \sqrt{-1}$, and $\theta_k, k = 1, \ldots, \frac{N}{2} - 1$ are randomly distributed between π and $-\pi$, then $h = \mathcal{F}^{-1}(H)$ is a cyclic all-pass sequence.

A transform employing the h and all its cyclic shifts as its basis can be easily implemented by cyclic correlation. If $x \in \Re^N$ is a vector of coefficients, the corresponding transform coefficients X can be obtained as

$$X = \mathcal{F}^{-1}(\mathcal{F}(x) \cdot \mathcal{F}(h)) \tag{8.2}$$

and the inverse transform can be obtained as

$$x = \mathcal{F}^{-1}(\mathcal{F}(X) \cdot \text{conj}(\mathcal{F}(h))). \tag{8.3}$$

Figure 8-10 shows the performance of the data hiding scheme for several test images undergoing JPEG (at various quality factors) and SPIHT compression (at different bit rates). From applying JPEG at quality factors of 75, 65, 55, and 50, respectively, it was found that the resulting images on an average were compressed to 1.35, 1.25, 1.15, 1.10, and 1.0 bpp, respectively. So, in the figure, the X-axis for both plots (JPEG and SPIHT) is an indication of the bit rate of the compression method employed.

Data hiding employed 8192 low-frequency magnitude-DFT coefficients. By subjecting various images to bitrate-N compression ($N = 1, \ldots, 5$, the X-axis), the average noise variances σ_ν^2 were estimated. The permitted distortion γ was chosen depending on the overall "activity" of the image. The measure of activity used was the MSE of the image after SPIHT compression at 1 bpp. The estimates of γ and σ_ν were used to obtain optimal values of embedding and detection parameters for each scenario as described in Section 5.2 where $\gamma^2 = \sigma_{X_n}^2$ and $\sigma_\nu^2 = \sigma_Z^2$.

8.2 Type III Hiding for Lossy Compression

Data compression is the most common application that any multimedia content will undergo. Therefore, optimal design of a watermarking method for the given compression is a very practical requirement. Given the quantization tables utilized by the compression scheme, one will know the exact compression noise that a stego signal will undergo. Hence, compression may be considered an attack in which the embedder has the ability to reduce its distorting effects on the stego signal.

As discussed in Chapter 5, the major advantage of quantization-based methods over additive schemes is that the former enables the hider to optimize the hiding rate at the given attack level, unlike the latter. Due to this property of type III methods, the embedding and detection parameters can be optimized in a way that takes into account compression distortion.

In this section, a type III data hiding scheme that makes use of the compression scheme's quantization characteristics is presented [94]. The method incorporates embedding quantization with the quantization of compression. Results show that joint embedding and compression has better payload and lower compression bit rates compared with independent compression and quantization. Hiding performance is evaluated under JPEG compression for the thresholding type of processing; however, the proposed methodology is trivially applicable to any lossy compression scheme for all types of postprocessing.

8.2.1   Joint Embedding and Compression

The motivation for modifying the embedder with respect to compression characteristics relies on the fact that the content creator, as the distributor, has the control over both watermarking and compression. Under this circumstance, an optimal system is the one that handles watermarking and compression jointly rather than considering them independent.

Considering watermarking and compression apart from each other may reduce the data hiding rate to remarkably low values or to zero. Among all possible cases, the worst occurs when the quantization step size specified by the compression scheme is much greater than Δ, the distance between the reconstruction points of the embedding quantizers. This may remove all the watermark and lead to zero hiding rates. Moreover, low hiding rates may not be avoided even in moderate or high bit-rate compression levels in such cases.

Embedding can be interpreted as introducing two forms of noise into the host signal, namely, the distortion due to embedding quantization and the processing distortion. The quantization involved in compression will round embedded watermark signal values to discrete quanta values. Therefore, the compression distortion, the difference between the watermarked signal and the quantized watermarked signal, is another source of noise that reduces the hiding rate. However, knowing the quantization characteristics in advance, the embedder can adjust its embedding distortion and processing distortion to lessen the effects of compression distortion. This requires the embedder to be modified in order to make comparisons between the watermarked signal and its quantized version, which will help to decide on the proper embedding and detection parameters. Using the *a priori* information on the compression, the embedder chooses among the (Δ, β) parameter pairs that maximize the data hiding rate. (Note that, as discussed in Chapter 5, for a permitted amount of embedding distortion, the information hider has infinitely many choices of embedding/detection parameter pairs.)

The information hiding system is outlined below:

$$\mathcal{W} : m \longrightarrow \mathbf{W},$$
$$\mathbf{S} = \mathcal{E}_Q(\mathbf{C}, \mathbf{W}) = \mathbf{S} + \mathbf{X_n},$$
$$\mathbf{Y} = \mathbf{S} + \mathbf{Q} + \mathbf{Z}, \tag{8.4}$$
$$\widehat{\mathbf{W}} = \mathcal{D}(\mathbf{Y}),$$
$$\mathcal{W}^{-1} : \widehat{\mathbf{W}} \longrightarrow \hat{m}$$

where $\mathbf{W}$ is the watermark signal corresponding to message index m, $\mathbf{C}$ represents the transformed cover signal coefficients, $\mathbf{X}_n$ is the type III codeword, $\mathbf{Q}$ is the quantization noise due to compression, and $\mathbf{Z}$ is the channel noise. Since quantization for lossy compression is generally performed in transform domain, embedder $\mathcal{E}_Q$ and the detector $\mathcal{D}$ operate on transform domain coefficients. The distortion introduced into $\mathbf{C}$ due to embedding, compression, and channel noise is measured using MSE distortion measures and is denoted by P_E, P_Q, and P_Z, respectively. The figure of merit used for evaluating the performance of the modified embedder is the normalized correlation between the embedded watermark signal and the extracted signal at varying ratios of distortion introduced by embedding and compression to channel noise distortion, $\frac{P_E+P_Q}{P_Z}$. Corresponding hiding rates are overestimated at a fixed $\frac{P_E+P_Q}{P_Z}$ through calculating the statistics of the Gaussian noise additive to the watermark signal vector so that the watermark signal vector and the extracted noised signal vector have the same correlation.

Comparing the joint and independent embedding/compression at the same distortion level of $P_Q + P_Z$, the hiding rate in the former will be higher as the mutual information between the $\mathbf{W}$ and $\widehat{\mathbf{W}}$ is higher, due to interrelated $\mathbf{X}_n$ and $\mathbf{Q}$. What is not so readily obvious is that better compression of the watermarked signal is possible when embedding is coordinated by the compression. As the embedder tries to minimize quantization noise by changing the embedded signal value with respect to its reconstruction value at the output of the quantizer, entropy of the quantized watermark signal decreases.

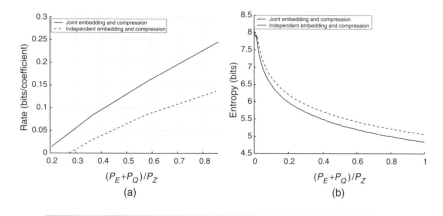

Figure 8-11 **(a)** Hiding rates for joint and independent embedding/compression. **(b)** Entropy of the quantized embedded signals.

Figure 8-11a displays the hiding rate vs robustness performance obtained for synthetically generated data using both joint and independent embedding/compression. The host signal $\mathbf{C}$ is assumed to be an *iid* Gaussian vector. For compression, a quantization step size of 6Δ is assumed for all coefficients. Fig. 8-11b displays the entropies for the watermarked signal after quantization for the same set of data. Joint embedding and compression has higher payload and provides a better compression of the watermarked signal when compared with independent embedding and compression.

8.2.2   Results for JPEG Compression

The method is implemented on a 256×256 sized test image where embedding is followed by a JPEG compression scheme [95]. A quality-factor concept introduced into the compression standard enables the provider to compress at various bit-rate values by scaling the built-in quantization tables. Transformed block coefficients are combined coherently into channels in which the first channel (00-channel) corresponds to DC coefficients and the rest of the 63 channels are for AC coefficients. The watermark signal is embedded into the first 9 low-frequency channels because the rest of

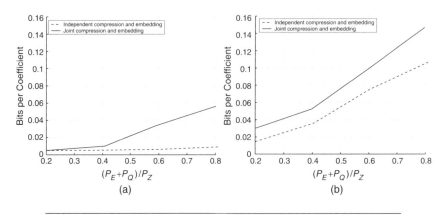

Figure 8-12 Hiding rates for 00-channel with compression at quality factors $(40 \leq P_E \leq 170)$ **(a)** JPEG-10 and **(b)** JPEG-50.

the channels go through a coarser quantization, which makes embedding extremely difficult.

The watermark signal embedded into the transformed image coefficients is an *iid* uniformly distributed vector of length 1024. This vector is embedded into the preselected low-frequency channels by the modified embedder making use of the quantization table for a particular quality factor. The attacker's intrusion is also modeled by an *iid* Gaussian noise vector of length 1024. Performance results are obtained for a range of $0.2 \leq \frac{P_E + P_Q}{P_Z} \leq 0.8$.

Figure 8-12a–b displays the improvement in 00-channel's hiding rate with joint embedding and compression, where embedding powers for JPEG-10 and JPEG-50 compression are restricted to be the same. Similarly, Fig. 8-13a–b displays the correctly detected number of bits among the embedded 1024 bits. Entropies of the watermarked images after quantization are displayed in Fig. 8-14a–b. The modified embedder contributes fewer bits per pixel increase to the compression bit rate of the sample image.

Although the modification on the embedder for joint embedding and compression is a simple one, the resulting benefits are twofold. Based on the *a priori* information on the compression, it becomes possible to achieve higher embedding rates by embedding with appropriate Δ and β values.

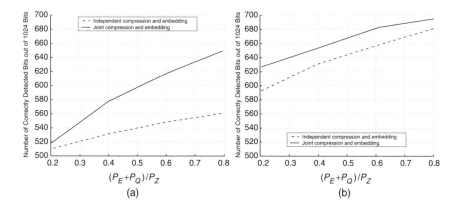

Figure 8-13   Number of correctly detected bits out of 1024 hidden bits for $(40 \leq P_E \leq 170)$ **(a)** JPEG-10 and **(b)** JPEG-50.

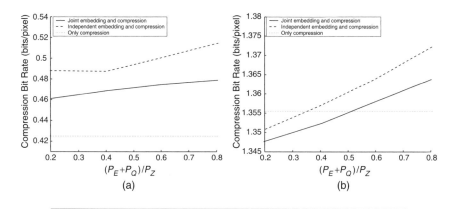

Figure 8-14   Entropy rates after quantization corresponding to **(a)** JPEG-10 and **(b)** JPEG-50.

Additionally, as the embedder aims to minimize quantization noise, the resultant embedded signal is more friendly to the quantization.

8.3   Watermarking for Ownership

Establishing ownership of books or blueprints has traditionally been done by obtaining copyright on their content, perhaps from the U.S. Copyright

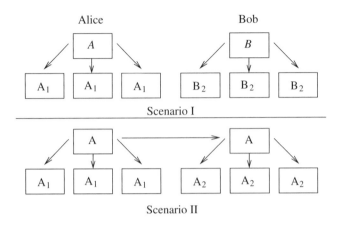

Figure 8-15 Scenarios in which existing copyright laws may be inadequate for resolving ownership. Scenario I (top): A and B are two similar photographs created by Alice and Bob. Scenario II (bottom): Alice (creator of A) does not want to obtain copyright.

Office. However, the nature of digital content makes traditional copyright mechanisms unsuitable for establishing ownership. Figure 8-15 depicts two typical scenarios, in which existing copyright mechanisms may be unsuitable for securing copyright of digital images. In scenario I, A and B represent two distinct but identical photographs created by Alice and Bob, respectively (both photographs may have been shot at the same place at different instances of time). Alice is responsible for circulating copies of her art as A_1. Meanwhile, Bob circulates his creation as B_2. Both Alice and Bob register their contents A and B with the U.S. Copyright Office.[3] If both A and B (and hence A_1 and B_2) look identical, Bob can claim that A and all A_1 are violations of his copyright, while Alice can claim that B all and B_2 are violations of her copyright. Obviously, this is not a desirable situation. In a second scenario, the photograph is created by Alice, who is not *interested* in obtaining a copyright. Bob may have received a copy of A (which Alice may have made freely available on her website), for which he promptly obtains a copyright and then circulates it as A_2. While it may

---

[3]To register a work of *visual art*, a completed application form, a nonrefundable filing fee of $30, and a nonreturnable deposit of the material to be registered are to be mailed to the Copyright Office. See http://www.loc.gov/copyright/reg.html for more details.

still be acceptable for Bob to claim ownership of all A_2 (circulated by Bob) it is definitely not ethical to let the copyright law enable Bob to claim ownership of the original A created by Alice. The key issue here (which cannot be determined by traditional copyright mechanisms) is to determine which copies *originate* from a *particular source*. *Watermarking the source* can effectively address this problem.

Digital watermarking [8] is a means of protecting multimedia content from intellectual piracy. It is achieved by imperceptibly modifying the original content to insert a "signature." The signature is extracted when necessary to show proof of ownership. It should be appreciated that watermarking is *just a tool* for protecting intellectual property rights, just as a lock is a tool for protecting a home. Like any tool, therefore, watermarking has to be used "intelligently." Just as a lock is useless in the hands of one who does not know how to *use* it, watermarking is useless without a proper "protocol" for using it. This section addresses the *need for a protocol* and *proposes a robust protocol* to make efficient use of watermarking. The proposed protocol takes the form of enhancements to the one suggested by Craver *et al.* [80]. In this paper, for the purpose of illustration, we assume that the original content is a digital image. However, the proposed protocol is equally applicable for video and audio signals as well.

8.3.1 Counterfeit Attacks on Watermarks

One of the primary issues to be addressed by watermarking methods is their ability to make a counterclaim *practically impossible*. A counterclaim arises from situations in which a pirate can use the *inadequacies of watermarking protocols* [80], [81], [96], [97] to "demonstrate" the presence of his/her "watermark" (a fake watermark or signature) in the actual original content.

Let $\mathbf{I}$ be the original (cover) image. A watermark embedding function E inserts a watermark $\mathbf{W}$ in the image $\mathbf{I}$ to generate the watermarked image $\hat{\mathbf{I}} = E(\mathbf{I}, \mathbf{W})$. The presence of the watermark $\mathbf{W}$ in an image $\tilde{\mathbf{I}}$ is checked by a detector D. Watermark detectors can be broadly classified into two categories. *Nonoblivious* detectors need the original image $\mathbf{I}$ to check for the presence of the signature $\mathbf{W}$ in $\tilde{\mathbf{I}}$. On the other hand, *oblivious* detectors [96], [98] *do not* require the original image. We shall term the output of the detector,

$$s_d = \begin{cases} D(\tilde{\mathbf{I}}, \mathbf{W}, \mathbf{I}) & \text{nonoblivious detector} \\ D(\tilde{\mathbf{I}}, \mathbf{W}) & \text{oblivious detector} \end{cases} \tag{8.5}$$

the *detection statistic*. The detection statistic is an indication of the *degree of certainty* with which the signature $\mathbf{W}$ is detected in the image $\tilde{\mathbf{I}}$.

Time stamping [99], [100] has been proposed as an *enhancement* to the security provided by watermarking to overcome the problems associated with counterclaims. In addition to watermarking, the creator can obtain a time stamp from a time stamping service (TSS). If the time stamp is obtained before the content is released to the public (before the *pirate* can obtain a time stamp on the content), nobody else can claim legitimate ownership of the content. However, time stamping has the disadvantage of requiring the *ongoing involvement* of a third party (i.e., the TSS). Moreover, there are some situations for which it does not provide acceptable solutions:

- Time stamping does not protect people who *do not* want to obtain a time stamp and/or watermark their content, like Alice in scenario II. If Bob is able to show a counterfeit signature in image A created by Alice, and if Alice has not obtained a time stamp, then Bob will be able to claim ownership of content created by Alice. Clearly, time stamping does not help in situations like this.
- Time stamping is not a solution for time-sensitive applications. The creator may not want to wait until he/she obtains a time stamp from a TSS. Therefore, it may be very difficult to use time stamping for securing live broadcasts.

However, we shall demonstrate that with a suitable protocol, which would lay some (very reasonable) restrictions on watermarking algorithms, these problems can be effectively addressed.

8.3.1.1 Freedom in Choosing

Let Alice be the creator of the original image $\mathbf{I}$. She embeds her signature $\mathbf{W}_A$ in $\mathbf{I}$ to obtain the watermarked image $\hat{\mathbf{I}}_A = E_A(\mathbf{I}, \mathbf{W}_A)$. The presence of her signature $\mathbf{W}_A$ in $\widehat{\mathbf{W}}_A$ or any image $\tilde{\mathbf{I}}_A$ *derived* from $\hat{\mathbf{I}}_A$ (or $\tilde{\mathbf{I}}_A = \hat{\mathbf{I}}_A + \mathbf{N}$) can be demonstrated, with a reasonably good degree of certainty, by obtaining a sufficiently high detection statistic,

$$d_{d_A} = D_A(\tilde{\mathbf{I}}_A, \mathbf{W}_A, \langle \mathbf{I} \rangle). \qquad (8.6)$$

In this equation, $\langle \mathbf{I} \rangle$ denotes that the original image $\mathbf{I}$ may or may not be used by the detector. The job of Bob, an aspiring pirate, is to demonstrate the presence of his (arbitrary) signature $\mathbf{W}_B$ in Alice's original image $\mathbf{I}$. In other words, Bob has to obtain a "large"

$$s_{d_B} = D_B(\mathbf{I}, \mathbf{W}_B, \langle \mathbf{I}_f \rangle) \qquad (8.7)$$

where $\mathbf{I}_f$ may be Bob's *fake original* image. Note that Bob is at liberty to *choose his own watermarking scheme* (E_B, D_B). If Bob has freedom in choosing his signature $\mathbf{W}_B$, he can fix some (E_B, D_B) and "construct" a signature $\mathbf{W}_B$ that yields a high detection statistic s_{d_B}. Note that even though Bob does not possess a copy of $\mathbf{I}$ (which is never released to the public by Alice), he does have $\hat{\mathbf{I}}_A$, which is "very close" to $\mathbf{I}$. Therefore, if Bob can choose a signature $\mathbf{W}_B$ such that $D_B(\hat{\mathbf{I}}_A, \mathbf{W}_B, \langle \mathbf{I}_f \rangle)$ is large, he is guaranteed that $s_{d_B} = D_B(\mathbf{I}, \mathbf{W}_B, \langle \mathbf{I}_f \rangle)$ will also be reasonably high.

If Bob does not have freedom in choosing his signature (for example, the signature may be *assigned* to him by a watermarking authority), he can still try to *tweak* the watermarking scheme (E_B, D_B) to obtain a high detection statistic.

Even if Bob's freedom in choosing both the signature and the watermarking scheme is curtailed, he could still tweak $\mathbf{I}_f$ (the fake original) to obtain a high detection statistic. Therefore, it is clear that a good watermarking protocol should address effective ways to limit all the "three degrees of freedom."

8.3.1.2 Detection Statistic

The detection statistic s_d [98] is a measure of degree of certainty with which the signature is detected. Typically, the signature $\mathbf{W}$ takes the form of a Gaussian or binary pseudo-random sequence (e.g., of length N) generated from a *key* $\mathcal{K}$. The watermark embedding and detection operations can therefore be written as

$$\hat{\mathbf{I}} = \mathcal{E}(\mathbf{I}, \mathbf{W}) \quad \widehat{\mathbf{W}} = \mathcal{D}(\tilde{\mathbf{I}}, \langle \mathbf{I} \rangle) \quad s_d = \frac{\mathbf{W}^T \widehat{\mathbf{W}}}{|\mathbf{W}| |\widehat{\mathbf{W}}|}. \tag{8.8}$$

In other words, the detection statistic $(-1 \le s_d \le 1)$ is a measure of (normalized) *inner product*, or normalized correlation of the embedded and detected signature sequences.

The normalized inner product of randomly generated signature sequences will also be random. More specifically, for large N, the distribution of the inner product will be Gaussian. Let x_i and y_i, $i = 1, \ldots, N$ be *iid* of variances $\sigma_x^2 = \sigma_y^2 = \frac{1}{N}$, and let $q_i = x_i y_i$. The inner product is calculated as $p = \sum_{i=1}^{N} q_i = \sum_{i=1}^{N} x_i y_i$. Since x_is and y_is are independent, the variance of q_i is $\sigma_q^2 = \sigma_x^2 \times \sigma_y^2 = \frac{1}{N} \times \frac{1}{N} = \frac{1}{N^2}$. Therefore, for large N, from central limit theorem [101], $p \sim \mathcal{N}[0, \frac{1}{N}]$.

If the creator (or pirate) has *absolutely no freedom* in choosing the signature, and if the detection statistic s_d obtained is, say, six times the standard deviation (if $s_d = 6\frac{1}{\sqrt{N}}$), then we could say that the signature is detected with a probability of error (or probability of false alarm) of less than $Q(6) \approx 1 \times 10^{-9}$, where $Q(x) = \frac{1}{\sqrt{2\pi}} \int_x^\infty e^{\frac{t^2}{2}} dt$ is related to the Gaussian error function. In other words, on average, only 1 out of 1×10^9 signatures chosen randomly can yield such a high correlation. Note that N is also the *degree of freedom* of the signature. Even if the detection statistic is unity, the signature is still detected with a nonvanishing false-alarm probability, given by $Q(\sqrt{N})$. *It is therefore very important to have a sufficiently large N to be able to yield "acceptable" levels of P_e.*

Any judge would be more than willing to rule in favor of detection of the signature if the probability of making a wrong decision is one in a million. In this case, $s_d = 5\frac{1}{\sqrt{N}}$ is more than acceptable. However, if the pirate can find a *loophole* in the watermarking protocol that enables a *search* for a suitable signature, then the pirate has to search for one million signatures (on average) before obtaining one that yields a satisfactory detection statistic.

One way to overcome this problem is to insist that the detection statistic be of the order of, say, $9\frac{1}{\sqrt{N}}$. This would imply that the pirate has to search through about 1×10^{19} signatures before being able to obtain one that yields satisfactory correlation. If a pirate could search 1×10^8 signatures in one second, then he/she would still need over 300 years to come up with a satisfactory signature! However, this restriction may make it considerably simpler for the pirate to *remove* the watermark by carefully planned attacks. After such attacks, the real owner may not be able to extract the signature with such a high degree of confidence (obtain a high detection statistic).

8.3.1.3 Fake Originals

As mentioned in Section 8.3.1.1, even if the watermarking scheme *and* the choice of signature are fixed, it may still be possible for a pirate to engineer a counterfeit attack *if the detection scheme is nonoblivious*. This would permit the pirate to create a *fake original* (cover) image, for which there are no restrictions! This problem can be solved if the detection method is *strictly* oblivious. But some geometric attacks on images like StirMark[4] [71] may be extremely difficult to overcome unless use the original image is permitted to *undo* the geometric distortions. Under this condition, the pirate may gain some freedom in *choosing* an algorithm for undoing the geometric distortions. Therefore, a good watermarking protocol should also fix (or *regulate*) the algorithms used for undoing the distortions introduced. However, the pirate still has *freedom in choosing the fake original that will be used by the fixed algorithm for undoing geometric distortions.* In other words, the pirate (Bob) can *engineer* a (fake) original that when used in conjunction with the fixed algorithm can "undo the distortion" in Alice's original image $\mathbf{I}$ (Bob would claim that the fake original he has created is the original and that Alice's image $\mathbf{I}$ is an image derived from his original) to show the presence of his watermark. Again, to engineer the attack, he has the image $\hat{\mathbf{I}}$, which is very close to $\mathbf{I}$ (Bob would try to tweak $\mathbf{I}_f$ to obtain a high value of $D_B(\hat{\mathbf{I}}, \mathbf{W}_B, \mathbf{I}_f)$).

8.3.1.4 Multiple Watermarks

Another possible counterfeit attack on watermarks is to create ambiguity in ownership due to the presence of multiple watermarks. Let $\mathbf{I}$ be the actual

---

[4]Free software available for download from http://www.cl.cam.ac.uk.

original created by Alice and $\hat{\mathbf{I}}_A$ the image containing Alice's watermark, which is available to Bob. Bob can create a new image $\hat{\mathbf{I}}_{AB}$ by embedding his signature in $\hat{\mathbf{I}}_A$. It is very possible for Bob to obtain a much higher detection statistic (of his signature) in $\hat{\mathbf{I}}_{AB}$ than Alice. In this case, what can prevent Bob from claiming ownership of $\hat{\mathbf{I}}_{AB}$? If Alice's watermarking scheme is robust, it might not be possible for Bob to show any copy of an image that is similar to $\mathbf{I}$ or $\hat{\mathbf{I}}_A$ or $\hat{\mathbf{I}}_{AB}$ and that *does not contain* Alice's signature. On the other hand, Alice's actual original $\mathbf{I}$, and $\hat{\mathbf{I}}_A$, will not contain Bob's signature. However, if Alice does not have a copy of the original (or $\hat{\mathbf{I}}_A$) in her possession, this problem may be difficult to resolve.

8.3.2 Watermarking Algorithms

Before we see how freedom in choosing can be curtailed, we need to understand the effect on watermarking algorithms. In this section we present a generalization of "conventional" watermarking algorithms. A few "unconventional" watermarking schemes that embed the watermark by introducing geometric distortions [90] or by modifying the histogram [89] have been proposed in literature. However, the main limitation of these algorithms is lack of robustness to commonly occurring attacks like lossy compression and/or drastically reduced degrees of freedom (N). The conventional algorithms rely on the assumption that if the image is altered significantly in the MSE sense, then the quality of the resultant image would be so poor that it would not warrant an ownership claim. Even though this assumption is not true, conventional algorithms can be *used effectively along with regulated algorithms for undoing geometric distortions/histogram modification, etc.*, as we shall see in the next section.

Typically, the watermark is inserted in some transform domain. Let $\mathbf{C_I} = \mathcal{T}(\mathbf{I})$, where $\mathcal{T}$ denotes some transform and $\mathbf{C_I}$ designates the transform coefficients of $\mathbf{I}$. Generally, only a subset of $\mathbf{C} \in \mathfrak{R}^N$ of $\mathbf{C_I}$ may be modified to embed the watermark, $\mathbf{W} \in \mathfrak{R}^N$, which is typically a pseudo-random binary or Gaussian sequence generated from a key $\mathcal{K}$. This general model does not preclude the possibility of $\mathcal{T}$ being an identity transform and the subset $\mathbf{C} = \mathbf{C_I}$. For enhanced robustness to intentional attacks, the transform (or its basis vectors) may also be generated from secret keys [102], [103]. Let $\mathbf{C_I} = \mathbf{C} \cup \bar{\mathbf{C}}$, where $\mathbf{C} \cap \bar{\mathbf{C}} = \emptyset$. The coefficients $\bar{\mathbf{C}}$ are unaffected by the watermarking process. The overall embedding operation may be expressed as

$$\mathbf{C_I} = \mathcal{T}(\mathbf{I}) \quad \mathbf{S} = \mathcal{E}(\mathbf{C}, \mathbf{W}) \quad \mathbf{S_I} = \mathbf{S} \cup \bar{\mathbf{C}} \quad \hat{\mathbf{I}} = \mathcal{T}^{-1}(\mathbf{S_I}). \tag{8.9}$$

Let $\tilde{\mathbf{I}} = \hat{\mathbf{I}} + \mathbf{N}$ be the image in which the presence of the watermark is tested. The detection operation can be expressed as

$$\mathbf{Y}_{\tilde{\mathbf{I}}} = \mathcal{T}(\tilde{\mathbf{I}}) \quad \widehat{\mathbf{W}} = \mathcal{D}(\mathbf{Y}) \quad s_d = \frac{\mathbf{W}^T \widehat{\mathbf{W}}}{|\mathbf{W}||\widehat{\mathbf{W}}|}. \tag{8.10}$$

Let $\mathbf{Z} = \mathbf{Y} - \mathbf{S}$ be the effect of $\mathbf{N}$ in the transform domain on the data-hidden coefficients $\mathbf{S}$. The watermarking algorithms that fit into the general model of Eqs. (8.9)–(8.10) are discussed in Chapters 3–5 as types I, II, and III.

8.3.3 Overcoming Attacks on Watermarks

Conventional watermarking methods introduce small modifications in the MSE sense ($|\mathbf{Y} - \mathbf{C}|^2$). Therefore, most attacks on conventional watermarks would rely on changing the image significantly in the MSE sense, without visually distorting the image. There are many ways to accomplish this—for example, scaling of pixel intensities, modifying the histogram, introducing small geometric distortions, etc. Similar attacks are also possible if the content is an audio signal instead of an image or video frame.

One way to survive geometric attacks like StirMark would require the watermarking method to *introduce* geometric distortions. Low-energy signatures introduced in the MSE sense show very good robustness to distortions of much higher energies due to *spreading gain*. Similarly, small modifications introduced into the geometric features can enable these features to withstand much larger attacks [90]. Let $\mathcal{G}(\mathbf{I})$ be a function of some geometric features of the image $\mathbf{I}$. The watermark can be introduced by *specifying* $\mathcal{G}(\hat{\mathbf{I}})$. However, we cannot expect such methods to be *robust to compression*. Just as small geometric distortions can modify the MSE significantly, small changes in MSE (such as those that might be introduced due to lossy compression) can change $\mathcal{G}(\mathbf{I})$ significantly. In this light it is not surprising that the watermarking method proposed by Rongen *et al.* [90] is robust to StirMark, but not very robust to lossy compression. Similarly, methods that specify the histogram [89] are not very resistant to compression.

One could still use conventional watermarking methods effectively if the primary ways by which the fake original can be moved away from the original in the MSE sense can be identified and suitable algorithms to *undo* the changes employed. For example, against attacks that modify the histogram, we could permit reshaping the histogram of the image in question to match the histogram of the original image before detecting the signature. Similarly a good algorithm for detecting *salient points* of the original image and those of the image in question may be used to re-warp the image (in which the signature is to be detected) so that the salient points match [104] before the signature is detected. Similar algorithms could also be used to overcome pixel scaling attacks. However, only *regulated* algorithms may be used for reshaping the histogram or identifying the salient points to re-warp the image or for rescaling the pixel values. As mentioned in Section 8.3.1.3, permitting freedom in choosing these algorithms would provide the pirate with additional degrees of freedom to engineer counterfeit attacks. Most oblivious watermarking methods proposed in the literature are not *strictly* oblivious. For strictly oblivious watermarking, the watermark detector may not even know the size of the original. The received image may have been resized, rotated, cropped, or undergone some histogram modification and probably some geometric distortion such as those introduced by StirMark. For example, the method proposed in [105] can reasonably survive RST attacks. This is achieved by embedding the watermark in a (significantly reduced degrees of freedom) RST invariant domain. However, it cannot survive RST *and cropping*. In order for it to do that, one might have to repeat the watermarks in many blocks of the image, thus reducing the degrees of freedom for the watermark further. Even if such a method survives RST and cropping, it may not be able to survive other forms of attacks. Reduced degrees of freedom imply lower separation between possible watermarks and, in general, lower robustness to attacks. As mentioned in Section 8.3.1.2, the maximum false alarm probability obtainable is $Q(\sqrt{N})$. Therefore, such schemes may not be able to *unambiguously establish ownership* in a court of law. In other words, a *strictly* oblivious watermarking method, capable of resolving ownership, may never be practical. In addition, if the creator does not maintain a copy of the original, the problem of multiple watermarks may be difficult to resolve (Section 8.3.1.4). In this light, insisting that the creator preserve an unwatermarked copy of his creation does not seem restrictive. It may be the only option. In fact, it has been shown in [106] that any method of resolving ownership with no reference to the original can never be unambiguous.

8.3.4 Restrictions on Choice of Signature

The restrictions on choice of signature, proposed in the watermarking literature, can be classified into those

  (1) issued by a watermarking authority (scheme I).
  (2) derived from a meaningful string [96] (scheme II), or
  (3) derived from the cover image [80] (scheme III).

Scheme I has the major disadvantage of requiring a watermarking authority in possession of all "secrets." The disadvantage of scheme II is that if the method of *obtaining the signature from the meaningful string* is fixed (as it should be—otherwise, the whole purpose is defeated), then it may be possible for the pirate (Bob) to "guess" the meaningful string used by Alice, thus reducing security. In addition, both schemes I and II suffer from the *fake original* problems discussed in Section 8.3.1.3. Moreover, neither of these schemes provide good solutions for the problem of multiple watermarks (Section 8.3.1.4).

In [80], Craver *et al.* suggested a novel idea (scheme III), which at one stroke solves the fake original problem, *the need for an agency* to issue signatures, and the multiple-watermarks problem. They suggested that the signature be obtained *from the original image itself*. The original image is hashed by a fixed hash function. The output is used as a seed for a *fixed random sequence generator* (RSG) to generate the signature. Tying the signature to the original image in an inextricable fashion goes a long way toward restricting the freedom available for the pirate to engineer counterfeit attacks. If the pirate tries to tweak the fake original, the signature also changes! The signature is obtained as $\mathbf{W}_A = \mathcal{G}(\mathcal{H}(\mathbf{I}))$, where $\mathcal{H}$ is a fixed hash function and $\mathcal{G}$ is a fixed RSG. More importantly, $\mathcal{H}(\mathbf{I}) \neq \mathcal{H}(\hat{\mathbf{I}})$. The watermarking scheme is nonoblivious and can therefore be described by

$$\mathbf{W}_A = \mathcal{G}(\mathcal{H}(\mathbf{I})) \quad \mathbf{S} = \mathbf{C} + \mathbf{W}_A \quad \widehat{\mathbf{W}}_A = \mathbf{Y} - \mathbf{C} \quad s_d = \frac{\mathbf{W}_A^T \widehat{\mathbf{W}}_A}{|\mathbf{W}_A||\widehat{\mathbf{W}}_A|}.$$
(8.11)

8.3.5   Attacking Scheme III (Craver's Protocol)

The attack, for Craver's protocol, rests on the fact that Bob can still *search* for a *combination* of a fake original and its corresponding signature. However, this is much more difficult than searching for a fake original alone, as is the case in schemes I and II, where the signature is fixed.

Bob, who has in his possession $\hat{\mathbf{I}}$ (or equivalently $\mathbf{S}$), could change $\hat{\mathbf{I}}$ significantly, in the MSE sense, while maintaining the visual similarity between the original $\hat{\mathbf{I}}$ and the resulting (modified) image $\mathbf{I}_m$. Let $\mathbf{I}_d$ be the difference image $\mathbf{I}_d = \mathbf{I}_m - \hat{\mathbf{I}}$ and $\mathbf{C}_d = \mathbf{C}_m - \mathbf{C}$. Even though the algorithms for undoing geometric distortions/histogram modifications/pixel rescaling would not permit Bob to move very far away from $\hat{\mathbf{I}}$, he should be able to introduce distortions such that the total power of $\mathbf{I}_d$ (or $\mathbf{C}_d$) is much larger than that of the signature $\mathbf{W}_A$ added by Alice. Alternately,

$$\sum_{i=1}^{N} C_d^2(k) \gg \sum_{i=1}^{N} (C(i) - S(i))^2.$$
(8.12)

Therefore, $\mathbf{C}_d = \mathbf{C}_m - \mathbf{S} \approx \mathbf{C}_m - \mathbf{C}_s$. The next step for Bob is to derive his "original" (fake original) image $\mathbf{I}_f$ from $\mathbf{I}_m$. Before we see how he can do that, note that the hash function $\mathcal{H}$ maps different images to (possibly) different seeds. For example, if all the images in the world were of size 256×256 and restricted to 8 bits per pixel, there would still be $2^{256 \times 256 \times 8}$ possible images. Though $\mathcal{H}$ would map the space of images to a (comparatively) very restricted "space" of seeds, this space should still be large enough to ensure that the probability of different signatures being correlated is very small. The case of two "obviously" different images having the same signature is not likely to create a problem. The problem arises only when images are similar. So, it is important that the (fixed) hash function generate different seeds, especially when the images are similar. This implies that the hash function would be required to "respond" to the LSBs of the image more than to its most significant bits. This works to Bob's advantage.

Bob could probably generate enough (different) signature sequences from the image $\mathbf{I}_m$ (or $\mathbf{C}_m$) just by tweaking one or two LSBs of the image pixels. But when he does this, the resulting image is still very close to $\mathbf{I}_m$. He could correlate every signature sequence obtained from modified versions of $\mathbf{I}_m$ with the fixed $\mathbf{C}_d$. Whenever a particular tweaking of the bits results in a signature sequence with satisfactory correlation with $\mathbf{C}_d$, he stops. He calls the resultant image $\mathbf{I}_f \approx \mathbf{I}_m$ as his (fake) original image. If $\mathbf{W_B}$ is the signature generated from $\mathbf{I}_f$, and $\mathbf{W}_b$ has a reasonable correlation with $\mathbf{S} - \mathbf{C}_f$, then it can also be expected to have high correlation with $\mathbf{C} - \mathbf{C}_f$. Thus Bob can demonstrate the presence of his signature in $\mathbf{I}$! Note that making $\mathbf{I}_f - \hat{\mathbf{I}}$ large swamps out the difference between $\mathbf{I}$ and $\hat{\mathbf{I}}$. After Bob's carefully planned series of attacks on Alice's watermark in $\mathbf{I}_m$ (en route to creation of $\mathbf{I}_f$), Alice may not be able to detect her signature in $\mathbf{I}_f$ with a high degree of certainty. Let us assume that Alice, using a very sophisticated watermarking method, manages to detect her signature in $\mathbf{I}_f$ with $P_e \approx 3 \times 10^{-7}$ (or $\rho_d = 5\frac{1}{\sqrt{N}}$). To obtain a comparable detection statistic of his signature in $\mathbf{I}$, Bob has to *search* roughly 3.3×10^6 sequences on average before obtaining a suitable signature. This is certainly feasible computationally.

8.3.6 Quasi-Oblivious Watermarking

We would like watermarking schemes to be able to resolve ownership unambiguously even in the face of different types of attacks like RST pixel scaling, histogram changes, imperceptible geometric attacks, lossy compression, etc. Furthermore, we would also like the watermarking scheme to be oblivious, as it would reduce *overheads necessary for detection* of the signature. However, an ideal oblivious watermarking scheme may never be practical. Firstly, as mentioned in Section 8.3.3, a strictly oblivious watermarking scheme, due to its vastly reduced degrees of freedom for the choice of signature, may not be able to yield satisfactorily low false-alarm probability P_e. Secondly, as shown by Qiao *et al.* [106], strictly oblivious watermarking may not be able to resolve ownership unambiguously.

The main necessity for an oblivious watermarking scheme arises out of the impracticality for the content creator to have access to his/her original,[5] to check for the presence of his/her signature in some content that may be present at a remote location. A typical scenario in which this might be required is when the content is rendered by a *compliant player* [102] (which can check for watermarks or *fingerprints*). The content provider would like to confirm that the content being rendered is indeed his/her content.

This problem can be solved by using two watermarks: one (which is not valid in court) for checking the presence of the creator's signature, and another, generated from the original, for unambiguously resolving ownership in court. However, the use of two independent signatures is a waste of bandwidth available for watermarking. A solution to this problem is to permit the content provider to have multiple keys. At least one of them, however, should be derived from the original. The others *have to be registered* with an appropriate authority. As a naive example, a content provider may use two keys: $\mathcal{K}_1$, a content-independent key registered with some authority, and $\mathcal{K}_2$ obtained from hashing the original. The key $\mathcal{K}_2$ is used to generate a ± 1 sequence of length K. The seed $\mathcal{K}_1$ is used to generate a "short" signature sequence $q \in \Re^{\frac{N}{K}}$. The signature sequence $\mathbf{W} \in \Re^N$ is obtained by the Kronecker product of q and the ± 1 sequence.[6] With the key $\mathcal{K}_1$, the *compliant player* can determine (by obtaining sufficiently high magnitudes of detection statistics of the short signature sequence q generated from $\mathcal{K}_1$) the identity of the content owner. If the content is being used in an unacceptable fashion, the content provider may need to take legal action and show unambiguously (by providing the pirated content along with the actual original in court) true ownership of the content.

---

[5]Especially for professional content creators who may have thousands of watermarked content.

[6]The watermarking scheme can also be seen as hiding a *content-dependent bit sequence* using a registered key $\mathcal{K}_1$.

8.3.7 Detection Statistic for Quasi-Oblivious Watermarking

Once the pirated content and the original are produced in court by the true owner, obtaining the detection statistic s_d, as in Eq. (8.10), may not be the best alternative. Making the detection statistic a *combination of the detection statistics obtained from the image in question and from the unwatermarked original* can go a long way toward rendering counterfeit attacks more difficult. As the watermarking scheme is oblivious, the watermark is detected *without subtracting* the original image. But the original image is still necessary because the seed is obtained from the original image, as in Craver's protocol. Further, we also need $P = \mathcal{D}(C)$. Let w, $\hat{w}$, and p represent normalized versions of W, $\widehat{W}$, and P, respectively, i.e., $w = \frac{W}{|W|}$. The detection statistic should be

$$s_d = w^T \hat{w} - b w^T p = w^T (\hat{w} - b w) \qquad (8.13)$$

where b is a scalar whose optimal choice shall be determined shortly.

Let $p_i, \hat{w}_i \sim \mathcal{N}[0, \frac{1}{N}]$, $i = 1, \ldots, N$. Obviously, as C and Y are *not independent*, p_i and $\hat{w}_i$, $i = 1, \ldots, N$ are not independent. Let $\rho = p^T \hat{w}$ and $u = \hat{w} - b p$. It can be easily shown that $u_i \sim \mathcal{N}\left[0, \sigma_u^2 = \frac{1 + b^2 - 2 b \rho}{N}\right]$ [101]. As we should try to minimize the false-alarm probability, we should try to minimize σ_u^2. The choice of b for minimizing σ_u is $b = \rho = p^T \hat{w}$. Therefore, the detection statistic should be

$$s_d = w^T (\hat{w} - b p) = w^T (\hat{w} - p^T \hat{w} p). \qquad (8.14)$$

Under this condition, $\sigma_u^2 = \frac{1 - \rho^2}{N}$.

The pirate, on the other hand, would try to maximize σ_u^2 to simplify his search. To do this he has to reduce ρ as much as possible by moving I_f far away from I. Bob, as earlier, would claim that I_f (C_f) is the original and would like to search for a signature W_f that would yield a high detection statistic with C. In other words, Bob's detection statistic is

$$s_{d_B} = w_B^T (p - p_f^T p p_f) \qquad (8.15)$$

where $\mathbf{P}_f = \mathcal{D}(\mathbf{C}_f)$ and $\mathbf{p}_f$ is the normalized version of $\mathbf{P}_f$. For type I oblivious methods, where $\mathbf{P} \equiv \mathbf{C}$ and $\mathbf{P}_f \equiv \mathbf{C}_f$, it may be very difficult to achieve low values of $\rho_B = \mathbf{p}_f^T \mathbf{p}$. The original and modified image[7] will be more similar than not similar, even in the MSE sense. As mentioned earlier, type III methods for watermarking are closer to type I. Therefore, it may be extremely difficult for Bob to obtain low values for ρ_B.

Note that for the genuine creator (Alice), the second term in Eq. (8.14) is not going to affect the signature extraction process in any way (the random signature generated from the image is very unlikely to yield a high correlation with the original image). However, we force the pirate to search for a signature that is not correlated with his fake original but at the same time is correlated with the actual original. This makes his search for a suitable signature extremely difficult. Let us assume that Alice obtains a detection statistic of $s_d = \frac{5}{\sqrt{N}}$ for her signature in Bob's image $\mathbf{I}_f$. For Bob to obtain a comparable detection statistic, he has to search for $\frac{1}{Q(5)} \approx 3 \times 10^6$ signatures with Craver's protocol. However, in the quasi-oblivious scheme suggested, Bob has to search an average of $Q(\frac{5}{\sqrt{1-\rho_b^2}})^{-1}$.

For $\rho_b = 0.6, 0.65, 0.7, 0.75$, and 0.8, Bob has to search for 4.9×10^9, 4.2×10^{10}, 7.9×10^{11}, 4.9×10^{13}, and 2.5×10^{16} signatures, respectively.

8.3.8 Suggested Protocol

Based on the arguments put forth in the previous sections, we suggest the following list of restrictions to be placed on watermarking methods, in order to make them resolve rightful ownership unambiguously; the overall protocol for watermark embedding and detection is shown in Figure 8-16. (In the figure, only the watermark-embedding and detection blocks will depend on the particular watermarking algorithm. The rest of the blocks are fixed [or regulated from time to time by the watermarking authority]):

(1) A prescribed algorithm for equalizing the histogram. The signature is added to the original content after equalizing its histogram. The histogram of the content in question is equalized

---

[7]The modified image after being subjected to regulated algorithms to undo distortions that might be introduced by the pirate.

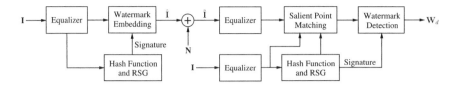

Figure 8-16   Watermark embedding and detection protocol.

(using the same equalizer) before performing detection of the signature.

(2) A prescribed algorithm for determining significant points and re-warping the image if necessary. For audio signals, the water-marked audio signal is resampled to ensure that salient points in the original (unwatermarked) and in the audio segment in question match. For instance, the algorithm presented in [104] by Ozer *et al.*, based on iterative partitioning and matching of "feature" points, helped in improving the false-alarm probability of detection of watermark in StirMarked images from as high as 10^{-2} to 10^{-50}.

(3) A prescribed algorithm for determining scale factors of pixel values/audio samples and rescaling and for equalizing the histogram prior to watermark detection.

(4) Fixed hash function $\mathcal{H}$ to be used. The hash function could be made computationally intensive to further discourage engineering of signatures. The hash function operates on the (histogram equalized) original content to produce the seed $\mathcal{H}(\mathbf{I})$. Other keys may be used in addition to the image-dependent key if they are registered with an appropriate authority.

(5) The seed $\mathcal{H}(\mathbf{I})$ is input to an RSG $\mathcal{G}$ to generate the signature sequence $\mathbf{W}_I$.

(6) $\mathbf{W}_N^d = \mathcal{G}(\mathcal{H}(\mathbf{I}), N, d)$ is the complete set of sequences that could be generated by $\mathcal{G}$. For a fixed $\mathbf{I}$ (or original content), the only parameters that can be changed are $N =$ the length of the sequence, and $d =$ the type of random sequence desired. For instance, d could take two options, Gaussian and uniform. Another useful option for d might be a random permutation of integers $1, \ldots, N$

(this may be used for reordering coefficients if the algorithm calls for it). No restriction is placed on the length N.

(7) Any decomposition of the original content can be used. If decompositions are generated from random sequences, only one from the set of possible sequences $\mathbf{W}_N^d$ can be used. If the watermarking algorithm calls for a random sequence (e.g., for reordering of coefficients) at any stage of the watermark embedding/extraction process, only random sequences $\mathbf{W}_N^d$ are permitted.

(8) The signature is to be extracted from the content in question without subtracting the original (or cover) content.

(9) The detection statistic should be the *weighted difference* between the statistics obtained from the content in question ($\tilde{s}_d$) and the statistics obtained from the original (s_{d_o}), using Eq. (8.14).

This proposed protocol does not limit itself only to methods in which the signature is detected by correlative processing. For example, in [107] some low-frequency DCT coefficients were scrambled by a random cyclic all-pass filter. The detection statistic was obtained by counting the difference between positive and negative coefficients. The only restrictions our proposal places on this method concerns how the seed is obtained and the corresponding random sequence to be used to generate the all-pass filter coefficients. To our knowledge, any existing oblivious detection watermarking method (with the exception of methods [89], [90] that introduce geometric distortions or modify the histogram to introduce the watermark) can be modified to meet the requirements of the proposed protocol. If the creator so desires, she may be able to resolve ownership unambiguously without the involvement of a third party if she uses only the key derived from hashing the original. In this case, the creator is also free to use nonoblivious data hiding.

8.3.9 An Example of a Watermarking Scheme

This section outlines a possible watermarking scheme for images (except for the choice of the decomposition employed, and the choice of coefficients to be modified for inserting the watermark, the proposed method is equally applicable for audio signals). The main purpose of this section is to illustrate with an example the influences of the proposed protocol

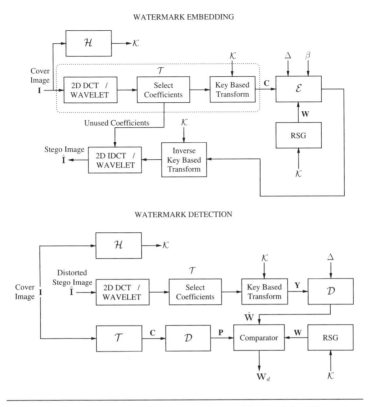

Figure 8-17 Block diagram of the watermark embedding and detection scheme.

in choosing parameters for the watermarking scheme. The block diagram of the scheme (embedding and detection) is shown in Figure 8-17. This block diagram may be considered a closer look at the shaded blocks in Figure 8-16.

In Figure 8-17, $\mathbf{I}$ represents the cover image after equalizing the histogram by the fixed equalizer. Perhaps, high-GTC transforms like DCT or wavelets are best suited for watermarking applications. As high-GTC transforms provide the most compact representation of the image, attacking DCT/wavelet coefficients for the purpose of watermark removal will most likely destroy the image. From the complete set of DCT/wavelet coefficients, we choose a low- to medium-frequency subset for watermarking purposes. The embedding and detection operators are type III.

For type II (and to a lesser extent, type III) systems, if the transform employed and quantization step size Δ are known, it is easy for a pirate to remove the signature completely without introducing significant distortion into the image. A truly secure watermarking scheme should be difficult to crack even if every step of the algorithm is public. In this case, the only "secret" should be the key $\mathcal{K}$ (which is derived from the original image using the hash function). The security can be vastly improved by using a key-based transform [102] before data embedding (and therefore before detection). In the proposed scheme, we use a simple key-based transform using cyclic all-pass filters as basis vectors [67], [98] (see Section 8.1.4).

The overall embedding operation is then as follows. The original image (after histogram equalization) undergoes DCT/wavelet transform, and selected low- to medium-frequency bands are utilized for data hiding. The selected coefficients are transformed by the key-based transform to obtain the coefficients $\mathbf{C}$ to be used for data embedding. The signature sequence $\mathbf{W}$ to be embedded into $\mathbf{C}$ may be obtained as a pseudo-random binary sequence using the prescribed RSG triggered by the key $\mathcal{K}$ (which in turn is derived from hashing the original image). The coefficients obtained after embedding, viz., $\mathbf{S}$, then undergo the inverse key–based transform to obtain the modified DCT/wavelet coefficients, which, together with the unmodified coefficients, are inverted to obtain the watermarked image, or the stego image.

For detection, the received image undergoes fixed algorithms for aligning geometric features and rescaling of pixel values/histogram equalization, resulting in image $\tilde{\mathbf{I}}$. The transformation $\mathcal{T}$ is performed on the received noisy image $\tilde{\mathbf{I}}$ to get the corresponding coefficients $\mathbf{Y}$. The detector function $\mathcal{D}$ extracts the noisy signature sequence $\widehat{\mathbf{W}}$, which, along with the signature sequence $\mathbf{W}$ (obtained from $\mathbf{I}$) and $\mathbf{P} = \mathcal{D}(\mathbf{C})$, is input to the comparator block. The comparator implements Eq (8.14) to obtain the detection statistic s_d. More details of the watermarking scheme and its performance can be found in [108].

Note that any permitted watermarking algorithm should have very little freedom in choosing arbitrarily defined parameters. For example, in this case the protocol may impose a condition that all watermarking algorithms should use the same Δ (which should be chosen after a lot of thought). A less restrictive (and probably more reasonable) rule could be that the value of Δ should have at least five significant digits—while the first digit may be chosen based on the design criteria, the following digits *should* be derived from the key $\mathcal{K}$, which itself is derived from the original.

CAE-CID Framework _____ under Varying Channel Noise

The optimal encoding and decoding described in Chapter 3.4 is achieved by the use of a shared collection of **U** sequences at the given channel noise level σ_Z^2. Consequently, when the channel noise level changes, successful operation can not be maintained due to the dependency on σ_Z^2. However, in CAE-CID framework, if encoder is aware of this change, reliable transmission can be restored by adjusting the input power without updating the shared collection of **U** sequences.

Each **U** sequence is an *iid* vector with the Gaussian marginal distribution, $U \sim \mathcal{N}(0, \sigma_X^2 + \sigma_C^2)$. Since both encoder and decoder are bound to use the same sequences (*i.e.* σ_X^2 and σ_C^2 are both fixed) and σ_X and σ_Z are related to each other due to Eq. (3.16) as

$$\sigma_X = \frac{P + \sigma_Z^2}{\sqrt{P}}, \tag{A.1}$$

encoder can adjust the input power in accordance with the new noise level $\hat{\sigma}_Z^2$. Using Eq. (A.1), the new input power $\hat{P}$ is found as

$$\sqrt{\hat{P}_{1,2}} = \frac{\sigma_X \pm \sqrt{\sigma_X^2 - 4\hat{\sigma}_Z^2}}{2} \tag{A.2}$$

where $\hat{P}_1$ and $\hat{P}_2$ are both valid choices only if $\sigma_X^2 - 4\hat{\sigma}_Z^2 \geq 0$ is satisfied. This requires $\sigma_X \geq 2\hat{\sigma}_Z$ as stated in Section 3.2.1, Eq. (3.20).

Consider $\hat{\sigma}_Z^2 = k\sigma_Z^2$, where $0 < k < \infty$, such that $0 < k < 1$ indicates a decrease in the channel noise and $1 < k < \infty$ indicates an increase compared to earlier state σ_Z^2. Since maximum communication rate is computed as $\frac{1}{2}\log_2\left(1 + \frac{P}{\sigma_Z^2}\right)$, Eq. (3.15), the new rate will change as a function of $\frac{\hat{P}}{\hat{\sigma}_Z^2}$, or equivalently $\frac{\sqrt{\hat{P}}}{\sqrt{k}\sigma_Z}$. Using Eq. (A.2) the change in $\frac{\sqrt{\hat{P}}}{\sqrt{k}\sigma_Z}$ with respect to $\frac{\sqrt{P}}{\sigma_Z}$ can be expressed as

$$r = \frac{\frac{\sqrt{\hat{P}}}{\sqrt{k}\sigma_Z}}{\frac{\sqrt{P}}{\sigma_Z}} = \frac{\frac{\sigma_X}{\sqrt{k}} \pm \sqrt{\frac{\sigma_X^2}{k} - 4\sigma_Z^2}}{\sqrt{P}}. \tag{A.3}$$

Since, for the given σ_Z^2, $\sqrt{P}$ needs to be a solution of Eq. (A.2), $\sqrt{P} = \frac{\sigma_X \pm \sqrt{\sigma_X^2 - 4\sigma_Z^2}}{2}$ will be true for one of the $\pm$. Then, the ratio given in Eq. (A.3) can be viewed as

$$r = \frac{\frac{\sqrt{\hat{P}}}{\sqrt{k}\sigma_Z}}{\frac{\sqrt{P}}{\sigma_Z}} = \frac{\frac{\sigma_X}{\sqrt{k}} \pm \sqrt{\frac{\sigma_X^2}{k} - 4\sigma_Z^2}}{\sigma_X \pm \sqrt{\sigma_X^2 - 4\sigma_Z^2}}. \tag{A.4}$$

Depending on the choice of $\hat{P}_1$ or $\hat{P}_2$ and k, the expression given in Eq. (A.3) will either be greater or smaller than 1. Therefore, when the channel noise changes from σ_Z^2 to $k\sigma_Z^2$, embedder and detector will be able to resume communication with the same set of $\mathbf{U}$ sequences at a lower or higher rate of $\frac{1}{2}\log_2\left(1 + r^2\frac{P}{\sigma_Z^2}\right)$ depending on the choice of input power, as given in Eq. (A.2), and k.

Statistics of _____

$\rho_{dep}|P$ and $d_{dep}|P$

The mean $m_{\rho*}$ of the random variable $\rho_{dep}|P$ can be computed through deriving the joint and marginal moments of W and $\widehat{W}$. The random variable $\widehat{W}$ is expressed in terms of Z_{eff} and W in Eq. (5.30), where W is a binary random variable with the density function $f_W(w) = \frac{1}{2}\delta(w-\frac{\Delta}{4})+\frac{1}{2}\delta(w+\frac{\Delta}{4})$. The pqth joint moment of W and $\widehat{W}$ is defined as

$$E[W^p\widehat{W}^q] = \int_{-\infty}^{\infty}\int_{-\infty}^{\infty} w^p\hat{w}^q f_{W,\widehat{W}}(w,\hat{w})\mathrm{d}w\mathrm{d}\hat{w}. \tag{B.1}$$

The joint pdf in the above integral can be expressed in terms of marginal and conditional pdf's, $f_{W,\widehat{W}}(w,\hat{w}) = f_{\widehat{W}}(\hat{w}|w_m)f_W(w)$, thus Eq. (B.1) can be written as

$$E[W^p\widehat{W}^q] = P(w=\frac{\Delta}{4})\int_{-\infty}^{\infty}\left(\frac{\Delta}{4}\right)^p \hat{w}^q f_{\widehat{W}}(\hat{w}|w=\frac{\Delta}{4})\mathrm{d}\hat{w}$$
$$+ P(w=-\frac{\Delta}{4})\int_{-\infty}^{\infty}\left(-\frac{\Delta}{4}\right)^p \hat{w}^q f_{\widehat{W}}(\hat{w}|w=-\frac{\Delta}{4})\mathrm{d}\hat{w}. \tag{B.2}$$

Since the expectation of a function of a random variable can be expressed in terms of the pdf of the random variable itself rather than of the function, $E[\widehat{W}] = \int_{-\infty}^{\infty} \hat{w}(z_{eff})f_{Z_{eff}}(z_{eff})\mathrm{d}z_{eff}$, and since all pdf's are assumed to be symmetric, Eq. (B.2) may be rewritten as

$$
\begin{aligned}
E[W^p\widehat{W}^q] &= \frac{1}{2}\sum_{i=-\infty}^{i=\infty}\int_{\frac{i\Delta}{2}}^{\frac{(i+1)\Delta}{2}} \left(\frac{\Delta}{4}\right)^p\left(\left(\frac{(2i+1)\Delta}{4}-z_{eff}\right)(-1)^i\right)^q \\
&\quad \times f_{Z_{eff}}(z_{eff})\mathrm{d}z_{eff} \\
&\quad + \frac{1}{2}\sum_{i=-\infty}^{i=\infty}\int_{\frac{i\Delta}{2}}^{\frac{(i+1)\Delta}{2}} \left(-\frac{\Delta}{4}\right)^p\left(\left(-\frac{(2i+1)\Delta}{4}+z_{eff}\right)(-1)^i\right)^q \\
&\quad \times f_{Z_{eff}}(z_{eff})\mathrm{d}z_{eff}, \\
&= \left(\frac{1}{2}+\frac{(-1)^{p+q}}{2}\right)\left(\frac{\Delta}{4}\right)^p 2 \\
&\quad \times \sum_{i=0}^{i=\infty}\int_{\frac{i\Delta}{2}}^{\frac{(i+1)\Delta}{2}} \left(\left(\frac{(2i+1)\Delta}{4}-z_{eff}\right)(-1)^i\right)^q f_{Z_{eff}}(z_{eff})\mathrm{d}z_{eff}, \\
&= \left(\frac{1}{2}+\frac{(-1)^{p+q}}{2}\right)\left(\frac{\Delta}{4}\right)^p R(q) \qquad\text{(B.3)}
\end{aligned}
$$

where $R(q)$ is as defined in Eq. (5.32). Hence, the joint moment of W and $\widehat{W}$ is generalized, based on Eq. (B.3), as

$$
E[W^p\widehat{W}^q] = \begin{cases} (\frac{\Delta}{4})^p R(q), & \text{if } p, q \text{ are both even or odd,} \\ 0, & \text{otherwise.} \end{cases} \qquad\text{(B.4)}
$$

Marginal moments of W are derived straightforwardly, due to the binary distribution, as

$$
E[W^p] = \begin{cases} 0, & \text{if } p \text{ is odd,} \\ (\frac{\Delta}{4})^p, & \text{if } p \text{ is even.} \end{cases} \qquad\text{(B.5)}
$$

The moments of the random variable $\widehat{W}$ depend on W and Z_{eff} through Eq. (5.30) and can be computed by using the properties employed in deriving

Eqs. (B.2) and (B.3) as

$$E\left[\widehat{W}^p\right] = P(W = \frac{\Delta}{4})E\left[\widehat{W}^p | w = \frac{\Delta}{4}\right] + P(W = -\frac{\Delta}{4})E\left[\widehat{W}^p | w = -\frac{\Delta}{4}\right],$$

$$= \frac{1}{2}\sum_{i=-\infty}^{i=\infty}\int_{\frac{i\Delta}{2}}^{\frac{(i+1)\Delta}{2}}\left(\left(\frac{(2i+1)\Delta}{4} - z_{\text{eff}}\right)(-1)^i\right)^p f_{Z_{\text{eff}}}(z_{\text{eff}})\mathrm{d}z_{\text{eff}}$$

$$+ \frac{1}{2}\sum_{i=-\infty}^{i=\infty}\int_{\frac{i\Delta}{2}}^{\frac{(i+1)\Delta}{2}}\left(\left(-\frac{(2i+1)\Delta}{4} + z_{\text{eff}}\right)(-1)^i\right)^p f_{Z_{\text{eff}}}(z_{\text{eff}})\mathrm{d}z_{\text{eff}},$$

$$= \left(\frac{1}{2} + \frac{1}{2}(-1)^p\right)2\sum_{i=0}^{i=\infty}\int_{\frac{i\Delta}{2}}^{\frac{(i+1)\Delta}{2}}\left(\left(\frac{(2i+1)\Delta}{4} - z_{\text{eff}}\right)(-1)^i\right)^p$$

$$\times f_{Z_{\text{eff}}}(z_{\text{eff}})\mathrm{d}z_{\text{eff}},$$

$$= \left(\frac{1}{2} + \frac{1}{2}(-1)^p\right)R(p). \tag{B.6}$$

Finally, $E\left[\widehat{W}^p\right]$ can be summarized as

$$E[\widehat{W}^p] = \begin{cases} 0, & \text{if } p \text{ is odd,} \\ R(p), & \text{if } p \text{ is even.} \end{cases} \tag{B.7}$$

Based on Eqs. (B.1)–(B.7), $m_{\rho*}$ is derived as

$$m_{\rho*} = \frac{E[W\widehat{W}]}{\sqrt{E[W^2]E[\widehat{W}^2]}},$$

$$= \frac{\frac{\Delta}{4}R(1)}{\sqrt{(\frac{\Delta}{4})^2 R(2)}},$$

$$= \frac{R(1)}{\sqrt{R(2)}}. \tag{B.8}$$

The variance $\sigma_{\rho*}^2$ is the variation of the correlation coefficient ρ^* around its mean $m_{\rho*}$ when $m_{\rho*}$ is estimated from N *iid* samples of $\widehat{\mathbf{W}}_m$

and $\mathbf{W}_m$. For the case when $\widehat{W}$ and W are from a bivariate Gaussian distribution, the variance is as given in [109]. However, when the samples are from non-Gaussian distributions, derivation of $\sigma_{\rho*}$ is not straightforward. Therefore, Monte Carlo simulations are performed to obtain the $\sigma^2_{\rho*}$ values for the considered N by computing the correlations between the embedded $\mathbf{W}_m$ and extracted $\widehat{\mathbf{W}}_m$ at the assumed WNR and then measuring the deviation from $m_{\rho*}$. However, for the minimum distance criterion, the corresponding variance values can be calculated in a straightforward manner.

For the minimum distance criterion, the statistics of $d_{dep}|P$ are computed in terms of the statistics of the random variable $\lambda = W^2 + \widehat{W}^2 - 2W\widehat{W}$.

When the noise level is very high, so that it can be considered uniformly distributed over all quantization intervals, W and $\widehat{W}$ become independent of each other, and $\widehat{W}$ is extracted as a uniformly distributed signal in $[-\frac{\Delta}{4}, \frac{\Delta}{4}]$. The mean $m_\lambda = E[\lambda]$, and the variance $\sigma^2_\lambda = E[\lambda^2] - m^2_\lambda$ of λ is calculated in terms of the moments

$$
\begin{aligned}
E[\lambda] &= \int_{-\infty}^{\infty} \int_{-\infty}^{\infty} (w^2 + \hat{w}^2 - 2w\hat{w}) f_{W,\widehat{W}}(w, \hat{w}) dw d\hat{w}, \\
&= \int_{-\infty}^{\infty} w^2 f_W(w) dw + \int_{-\infty}^{\infty} \hat{w}^2 f_{\widehat{W}}(\hat{w}) d\hat{w} \\
&\quad - 2 \int_{-\infty}^{\infty} w f_W(w) dw \int_{-\infty}^{\infty} \hat{w} f_{\widehat{W}}(\hat{w}) d\hat{w}, \\
&= Var[W] + Var[\widehat{W}], \\
&= \frac{\Delta^2}{16} + \frac{\Delta^2}{48} = \frac{\Delta^2}{12},
\end{aligned} \tag{B.9}
$$

$$
\begin{aligned}
E[\lambda^2] &= \int_{-\infty}^{\infty} \int_{-\infty}^{\infty} (w^2 + \hat{w}^2 - 2w\hat{w})^2 f_{W,\widehat{W}}(w, \hat{w}) dw d\hat{w}, \\
&= \int_{-\infty}^{\infty} w^4 f_W(w) dw + \int_{-\infty}^{\infty} \hat{w}^4 f_{\widehat{W}}(\hat{w}) d\hat{w} \\
&\quad + 6 \int_{-\infty}^{\infty} w^2 f_W(w) dw \int_{-\infty}^{\infty} \hat{w}^2 f_{\widehat{W}}(\hat{w}) d\hat{w}, \\
&= \frac{\Delta^4}{2^8} + \frac{\Delta^4}{2^8 5} + \frac{\Delta^4}{2^7} = \frac{\Delta^4}{80},
\end{aligned} \tag{B.10}
$$

as $m_\lambda = \frac{\Delta}{12}$ and

$$\sigma_\lambda^2 = \frac{\Delta^4}{80} - \left(\frac{\Delta^2}{12}\right)^2 = \frac{\Delta^4}{180}. \tag{B.11}$$

When W and $\widehat{W}$ are dependent on each other, the statistics of $d_{dep}|P$ can be similarly computed in terms of the individual and joint moments of W and $\widehat{W}$, Eqs. (B.4)–(B.7). Consequently the mean m_{d*} and the variance σ_{d*}^2 are computed as

$$
\begin{aligned}
m_{d*} &= E\left[\frac{1}{N}\sum_{l=1}^{l=N}\left(W^2 + \widehat{W}^2 - 2W\widehat{W}\right)\right], \\
&= E[W^2] + E[\widehat{W}^2] - 2E[W\widehat{W}], \\
&= \left(\frac{\Delta}{4}\right)^2 + R(2) - \left(\frac{\Delta}{4}\right)R(1),
\end{aligned}
\tag{B.12}
$$

$$
\begin{aligned}
\sigma_{d*}^2 &= E\left[\left(\frac{1}{N}\sum_{l=1}^{l=N}\left(W^2 + \widehat{W}^2 - 2W\widehat{W}\right)\right)^2\right] - m_{d*}^2, \\
&= \frac{1}{N}\left(E[W^4] + E[\widehat{W}^4] + 6E[W^2\widehat{W}^2] - 4E[W^3\widehat{W}] - 4E[W\widehat{W}^3]\right) \\
&\quad + \frac{N-1}{N}\Big(E[W^2]^2 + E[\widehat{W}^2]^2 - 4E[W^2]E[W\widehat{W}] - 4E[\widehat{W}^2]E[W\widehat{W}] \\
&\quad + 2E[W^2]E[\widehat{W}^2] + 4E[W\widehat{W}]^2\Big) - m_{d*}^2, \\
&= \frac{1}{N}\left(\left(\frac{\Delta}{4}\right)^4 + R(4) + 6\left(\frac{\Delta}{4}\right)^2 R(2) - 4\left(\frac{\Delta}{4}\right)^3 R(1) - 4\left(\frac{\Delta}{4}\right)R(3)\right) \\
&\quad + \frac{N-1}{N}\left(\left(\frac{\Delta}{4}\right)^4 + R(2)^2 - 4\left(\frac{\Delta}{4}\right)^3 R(1) - 4\frac{\Delta}{4}R(2)R(1)\right. \\
&\quad \left. + 2\left(\frac{\Delta}{4}\right)^2 R(2) + 4\left(\frac{\Delta}{4}\right)^2 R(1)^2\right) - m_{d*}^2.
\end{aligned}
\tag{B.13}
$$

Mathematical Proofs_____

C.1   Proof of Eq. (7.7)

$$h(n) = \sum_{k=0}^{N-1} e^{j(\frac{2\pi kn}{N}+\phi_k)} \qquad f(n) = \sum_{k=0}^{N-1} a_k e^{j(\frac{2\pi kn}{N}+\theta_k)}$$

for $n = 0 \cdots N - 1$. From Eq. (7.6),

$$
\begin{aligned}
\varepsilon &= \sum_{n=0}^{N-1}\sum_{l=0}^{N-1}\sum_{k=0}^{N-1} \left[e^{j(\frac{2\pi kn}{N}+\phi_k)} - a_k e^{j(\frac{2\pi kn}{N}+\theta_k)} \right] \times \left[e^{-j(\frac{2\pi ln}{N}+\phi_l)} - a_l e^{-j(\frac{2\pi ln}{N}+\theta_l)} \right] \\
&= \sum_{k=0}^{N-1}\sum_{l=0}^{N-1} \left(\sum_{n=0}^{N-1} \left[e^{j(\frac{2\pi(k-l)n}{N})} e^{j(\phi_k-\phi_l)} - a_l e^{j(\frac{2\pi(k-l)n}{N})} e^{j(\phi_k-\theta_l)} \right. \right.\\
&\qquad \left. \left. - a_k e^{j(\frac{2\pi(k-l)n}{N})} e^{j(\theta_k-\phi_l)} + a_k a_l e^{j(\frac{2\pi(k-l)n}{N})} e^{j(\theta_k-\theta_l)} \right] \right).
\end{aligned}
\tag{C.1}
$$

Using the identity

$$\sum_{n=0}^{N-1} e^{j(\frac{2\pi(k-l)n}{N})} = \begin{cases} N & \text{for } k = l \\ 0 & \text{otherwise} \end{cases}, \tag{C.2}$$

Eq. (C.1) reduces to

$$\varepsilon = N \left[N - 2 \sum_{k=0}^{N-1} a_k \cos(\phi_k - \theta_k) + \sum_{k=0}^{N-1} a_k^2 \right]. \tag{C.3}$$

C.2   Proof of Eq. (7.10)

Given that $h \in \Re^N$, $H = \mathcal{F}_N(h)$, and $h_e(n) = h(2n)$ for $n = 0, \ldots, \frac{N}{2} - 1$, and $H_e = \mathcal{F}_{N/2}(h_e)$, we need to show

$$H_e(l) = \frac{H(l) + H(l + \frac{N}{2})}{2}, \quad l = 0 \cdots \frac{N}{2} - 1. \tag{C.4}$$

$$
\begin{aligned}
H_e(l) &= \sum_{n=0}^{\frac{N}{2}-1} h(2n) \exp\left(\frac{-j2\pi nl}{\frac{N}{2}}\right) \tag{C.5}\\
&= \sum_{n=0}^{\frac{N}{2}-1} \frac{1}{N} \sum_{k=0}^{N-1} H(k) \exp\left(\frac{j4\pi nk}{N}\right) \exp\left(\frac{-j4\pi nl}{N}\right)\\
&= \frac{1}{N} \sum_{k=0}^{N-1} H(k) \sum_{n=0}^{\frac{N}{2}-1} \exp\left(\frac{j4\pi n(k-l)}{N}\right)\\
&= \frac{H(l) + H(l + \frac{N}{2})}{2}, \quad l = 0 \cdots \frac{N}{2} - 1.
\end{aligned}
$$

Bibliography

1. F. L. Bauer. *Decrypted Secrets: Methods and Maxims of Cryptology*. New York: Springer Verlag, 1997.
2. David Kahn. *The Codebreakers*. New York: Macmillan, 1967.
3. Jagadguru Swami Shri Bharati Krishna Tirthaji Maharaja. *Vedic Mathematics*. Motilal Banarsidass, Delhi, 1988.
4. K. Mantusi and K. Tanaka. Video steganography: How to secretly embed a signature in a picture? In *Proc. of IMA Intellectual Property Project, Interactive Multimedia Association*, pages 263–272, 1994.
5. J. Zhao, E. Koch, and C. Luo. In business today and tomorrow. *Communications of the ACM*, 41(7):66–72, 1998.
6. A. H. Tewfik. White paper on data embedding. *Media Annotation and Copyright Protection Product Line, http://www.cognicity.com*, 1998.
7. J. Brassil, S. Low, N. Maxemchuk, and L. Ó. Gorman. Electronic marking and identification techniques to discourage document copying. *IEEE Journal on Selected Areas in Communication*, 13(8):1495–1504, 1995.
8. I. J. Cox, J. Kilian, T. Leighton, and T. Shamoon. Secure spread spectrum watermarking for multimedia. *IEEE Transactions on Image Processing*, 6(12):1673–1687, 1997.
9. B. Chen and G. Wornell. Preprocessed and postprocessed quantization index modulation methods for digital watermarking. In *Proc SPIE: Security and Watermarking of Multimedia Contents II*, volume 3971, pages 48–59, 2000.
10. I. J. Cox, M. L. Miller, and A. L. McKellips. Watermarking as communication with side information. *Proc. of IEEE*, 87:1127–1141, 1999.
11. J. Chou, S. S. Pradhan, L. E. Ghaoui, and K. Ramchandran. On the duality between data hiding and distributed source coding. In *Proc. of 33rd Annual Asilomar conference on Signals, Systems, and Computers*, 1999.
12. R. J. Barron, B. Chen, and G. W. Wornell. The duality between information embedding source coding with side information and its implications— applications. *IEEE Transactions on Information Theory*, 49(5):1159–1180, 2003.
13. W. Bender, D. Gruhl, N. Morimoto, and A. Lu. Techniques for data hiding. *IBM Systems Journal*, 35(3-4):313–336, 1996.

14. I. J. Cox, J. Kilian, T. Leighton, and T. Shamoon. A secure, robust, watermark for multimedia. In *Proc. of 1st int. Information Hiding Workshop*, pages 185–206, 1996.

15. J. R. Smith and B. O. Comiskey. Modulation and information hiding in images. In *Proc. of 1st int. Information Hiding Workshop*, pages 207–226, 1996.

16. F. Hartung and B. Girod. Digital watermarking of raw compressed video. In *Proc. of European Conference of Advanced Imaging and Network Technologies*, 1996.

17. J. R. Hernandez, F. Perez-Gonzalez, J. M. Rodriguez, and G. Nieto. Performance analysis of a 2-d multipulse amplitude modulation scheme for data hiding and watermarking of still images. *IEEE J. Select. Areas Commun.*, 16(4):510–524, 1998.

18. M. Costa. Writing on dirty paper. *IEEE Transactions on Information Theory*, 29:439–441, 1983.

19. B. Chen and G. W. Wornell. Provably robust digital watermarking. In *Proc SPIE: Multimedia Systems and Applications*, volume 3845, 1998.

20. B. Chen and G. W. Wornell. Quantization index modulation: A class of provably good methods for digital watermarking and information embedding. *IEEE Transactions on Information Theory*, 47(4):1423–1443, May 2001.

21. M. Ramkumar and A. N. Akansu. Self-noise suppression schemes for blind image steganography. In *Proc. SPIE International Workshop on Voice, Video and Data Communication, Multimedia Applications*, volume 3845, September 1999.

22. J. J. Eggers, J. K. Su, and B. Girod. A blind watermarking scheme based on structured codebooks. *IEE Colloq. Secure Images and Image Authentication*, 4:1–6, April 2000.

23. J. Chou, S. S. Pradhan, L. E. Ghaoui, and K. Ramchandran. A robust optimization solution to the data hiding problem using distributed source coding principles. In *Proc. SPIE: Image and Video Communications and Processing*, volume 3974, 2000.

24. R. Zamir, S. Shamai, and U. Erez. Nested linear/lattice codes for structured multiterminal binning. *IEEE Transactions on Information Theory*, 48(5):1250–1276, 2002.

25. B. Chen and G. W. Wornell. Digital watermarking and information embedding using dither modulation. In *IEEE Second Workshop on Multimedia Signal Processing*, pages 273–278, 1998.

26. B. Chen and G. W. Wornell. Dither modulation: A new approach to digital watermarking and information embedding. In *Proc. of SPIE: Security and Watermarking of Multimedia Contents*, volume 3657, pages 342–353, 1999.

27. K. Tanaka, Y. Nakamura, and K. Matsui. Embedding secret information into a dithered multi-level image. In *Proc. of IEEE International Conference On Image Processing*, pages 216–220, 1990.

28. R. G. van Schyndel, A. Z. Tirkel, and C. F. Osborne. A digital watermark. In *Proc. of IEEE International Conference On Image Processing*, volume 2, pages 86–90, 1994.

29. G. Caronni. Assuring ownership rights for digital images. In *Proc. of Reliable IT Systems*, volume VIS-95. Vieweg Publishing Company, 1995.

30. M. D. Swanson, B. Zhu, and A. H. Tewfik. Data hiding for video-in-demand. In *Proc. of IEEE International Conference On Image Processing*, volume 2, pages 676–679, 1997.

31. H.-J. M. Wang, P.-C. Su, and C.-C. J. Kuo. Wavelet-based digital image watermarking. *Optics Express*, 3(12):491–496, December 1998.

32. M. Wu and B. Liu. Watermarking for image authentication. In *Proc. of IEEE International Conference On Image Processing*, volume 2, pages 437–441, 1998.

33. B. Chen and G. Wornell. Achievable performance of digital watermarking systems. In *Proc. of IEEE Int. Conference on Multimedia Computing and Systems*, volume 1, pages 13–18, 1999.

34. J. J. Eggers. *Information Embedding and Digital Watermarking as Communication with Side Information*. Ph.D. thesis, Lehrstuhl fur Nachrichtentechnik I, Universitat Erlangen-Nurnberg, Erlangen, Germany, 2001.

35. J. J. Eggers, R. Bauml, R. Tzschoppe, and B. Girod. Scalar costa scheme for information embedding. *IEEE Transactions on Signal Processing*, 51(4):1003–1019, 2003.

36. F. Perez-Gonzalez, F. Balado, and J. R. Hernandez Martin. Performance analysis of existing and new methods for data hiding with known-host information in additive channels. *IEEE Transactions on Signal Processing*, 51(4):960–980, 2003.

37. H. Malvar and D. A. F. Florencio. Improved spread spectrum: a new modulation for robust watermarking. *IEEE Transactions on Signal Processing*, 51(4):898–905, 2003.

38. C. E. Shannon. Channels with side information at the transmitter. *IBM Journal of Research and Development*, 2:289–293, 1958.

39. A. V. Kusnetsov and B. S. Tsybakov. Coding in a memory with defective cells. *Translation from Problemy Peredachi Informatsi*, 10:52–60, 1974.

40. S. I. Gelfand and M. S. Pinsker. Coding for channel with random parameters. *Problems of Control and Information Theory*, 9(1):19–31, 1980.

41. C. Heegard and A. A. El Gamal. On the capacity of computer memory with defects. *IEEE Transactions on Information Theory*, 29:731–739, 1983.

42. P. Moulin and J. A. O'Sullivan. Information-theoretic analysis of information hiding. *IEEE Transactions on Information Theory*, 49:563–593, March 2003.

43. A. S. Cohen and A. Lapidoth. The gaussian watermarking game. *IEEE Transactions on Information Theory*, 48:1639–1667, June 2002.

44. T. M. Cover and J. A. Thomas. *Elements of Information Theory, Second Edition*. New York: John-Wiley & Sons Inc., 1991.

45. U. Erez, S. Shamai, and R. Zamir. Capacity and lattice-stratergies for cancelling known interference. In *Proc. of Int. Symp. Information Theory and Its Application*, pages 681–684, 2000.

46. H. T. Sencar. *Oblivious Data Hiding: A Practical Approach*, Ph.D. thesis, New Jersey Institute of Technology, Newark, NJ, 2004.

47. H. T. Sencar, M. Ramkumar, and A. N. Akansu. Efficient codebook structures for practical information hiding systems. In *Proc. of CISS*, March 2001.

48. M. Ramkumar. *Data Hiding in Multimedia-Theory and Applications*. PhD thesis, New Jersey Institute of Technology, Newark, NJ, 2000.

49. H. T. Sencar, M. Ramkumar, and A. N. Akansu. A new perspective for embedding-detection methods with distortion compensation and thresholding processing techniques. In *Proc. of IEEE-ICIP Conference*, 2003.

50. M. Ramkumar and A. N. Akansu. Capacity estimates for data hiding in compressed images. *IEEE Transaction on Image Processing*, 10(8):1252–1263, 2001.

51. M. Ramkumar and A. N. Akansu. Information theoretic bounds for data hiding in compressed images. In *IEEE Second Workshop on Multimedia Signal Processing*, pages 267–272, 1998.

52. J. Bloom, M. Miller, and I. Cox. *Digital Watermarking: Principles and Practice*. San Francisco: Morgan Kaufmann, 2001.

53. R. B. Wolfgang, C. I. Podilchuk, and E. J. Delp. The effect of matching watermark and compression transforms in compressed color images. In *IEEE International Conference on Image Processing*, volume 1, pages 440–443, 1998.

54. A. Said and W. A. Pearlman. A new fast and efficient implementation of an image codec based on set partitioning in hierarchical trees. *IEEE Transactions on Circuits and Systems for Video Technology*, 6:243–250, 1996.

55. H. -J. M. Wang, P. -C. Su, and C. -C. J. Kuo. Wavelet based digital image watermarking. *Optics Express*, 3(12):491–496, 1998.

56. J. R. Smith and B. O. Comiskey. Modulation and information hiding in image. In *Workshop on Information Hiding, University of Cambridge, UK*, pages 463–470, 1996.

57. S. D. Servetto, C. I. Podilchuk, and K. Ramachandran. Capacity issues in digital watermarking. In *IEEE International Conference on Image Processing*, volume 1, pages 445–448, 1998.

58. J. R. Hernandez, F. Perez-Gonzalez, J. M. Rodriguez, and G. Nieto. Performance analysis of a 2-d multipulse amplitude modulation scheme for data hiding and watermarking of still images. *IEEE Journal on Selected Areas in Communications*, 16(4):510–524, 1998.

59. A. N. Akansu and R. A. Haddad. *Multiresolution Signal Decomposition: Transforms, Subbands and Wavelets.* New York: Academic Press Inc., 1992.

60. H. T. Sencar, M. Ramkumar, and A. N. Akansu. An overview of scalar quantization based data hiding methods. *Submitted to IEEE Transactions on Signal Processing.*

61. R. G. Gray and T. M. Stockham. Dithered quantizers. *IEEE Transactions on Information Theory*, 39(3):805–812, 1993.

62. J. J. Eggers, J. K. Su, and B. Girod. Public key watermarking by eigenvectors of linear transforms. In *European Signal Processing Conference (EUSIPCO 2000)*, September 2000.

63. H. T. Sencar, M. Ramkumar, and A. N. Akansu. Multiple codebook information hiding. In *Proc. of CISS*, 2002.

64. H. T. Sencar, M. Ramkumar, and A. N. Akansu. Multiple codebook information hiding based on minimum distortion criterion. In *Proc. of CISS*, March 2003.

65. H. T. Sencar, M. Ramkumar, and A. N. Akansu. An analysis of quantization based embedding-detection techniques with very large and small embedding signal sizes. Submitted to *IEEE Transactions on Signal Processing.*

66. D. S. Watkins. *Fundamentals of Matrix Computations.* New York: John Wiley & Sons, 1991.

67. Mahalingam Ramkumar, G. V. Anand, and Ali N. Akansu. On the implementation of the 2-band cyclic filter banks. In *Proc. of IEEE International Symposium Circuits and Systems*, volume 3, pages 520–523, 1999.

68. S. B. Wicker. *Error Control Systems for Digital Communication and Storage.* Englewood Cliffs, NJ: Prentice Hall, 1995.

69. F. A. P. Petitcolas, R. J. Anderson, and M. G. Kuhn. Attacks on copyright marking systems. In *Second Workshop on Information Hiding, Lecture Notes in Computer Science*, volume 1525, pages 218–238, 1998.

70. I. J. Cox and J. P. Linnartz. Some general methods for tampering with watermarks. *IEEE Journal on Selected Areas in Communications*, 16(4):587–593, 1998.

71. M. Kutter and F. A. P. Petitcolas. A fair benchmark for image watermarking systems. In *Proc. of SPIE Security and Watermarking of Multimedia Contents*, volume 3657, pages 226–239, 1999.

72. M. Kutter. Watermarking resistent to translation, rotation, and scaling. In *Proc. SPIE Multimedia Systems and Applications*, volume 3528, pages 423–431, November 1998.

73. E. T. Lin and E. J. Delp. Temporal synchronization in video watermarking. In *Proc. SPIE Security and Watermarking of Multimedia Contents IV*, volume 4675, January 2002.

74. H. T. Sencar, M. Ramkumar, and A. N. Akansu. A robust type III data hiding technique against cropping and resizing attacks. In *IEEE International Symposium Circuits and Systems*, volume 4, pages 3449–3452, 2002.

75. F. Hartung, J. K. Su, and B. Girod. Spread spectrum watermarking: Malicious attacks and counterattacks. In *Proc. of SPIE Security and Watermarking of Multimedia Contents*, pages 147–158, 1999.

76. S. Craver, N. D. Memon, B.-L. Yeo, and M. M. Yeung. Can invisible watermarks resolve rightful ownerships? In *Storage and Retrieval for Image and Video Databases (SPIE)*, pages 310–321, 1997.

77. S. Voloshynovskiy, S. Pereira, V. Iquise, and T. Pun. Attack modelling: Towards a second generation benchmark. *Signal Processing*, 81(6):1177–1214, June 2001. Special Issue: Information Theoretic Issues in Digital Watermarking, 2001. V. Cappellini, M. Barni, F. Bartolini, Eds.

78. S. Voloshynovskiy, S. Pereira, T. Pun, J. Eggers, and J. Su. Attacks on digital watermarks: Classification, estimation-based attacks and benchmarks. *IEEE Communications Magazine* (Special Issue on Digital watermarking for copyright protection: a communications perspective), 39(8):118–127, 2001.

79. S. Pereira, S. Voloshynovskiy, M. Madueño, S. Marchand-Maillet, and T. Pun. Second generation benchmarking and application oriented evaluation. In *Information Hiding Workshop*, Pittsburgh, Apr 2001.

80. S. Craver, N. Memon, B-L. Yeo, and M. M. Yeung. Can invisible watermarks resolve rightful ownerships. In *IS & T/ SPIE Electronic Imaging: Human Vision and Electronic Imaging*, volume 3022, pages 310–321, 1997.

81. S. Craver, N. Memon, B.-L. Yeo, and M. M. Yeung. Resolving rightful ownerships with invisible watermarking techniques: Limitations, attacks, and implications. *IEEE Journal on Selected Areas in Communications*, 16(4):573–586, 1998.

82. M. Ramkumar and A. N. Akansu. Image watermarks and counterfeit attacks: Some problems and solutions. In *Content Security and Data Hiding in Digital Media*, pages 102–112, 1999.

83. M. Ramkumar and A. N. Akansu. Robust protocols for proving ownership of multimedia content. *IEEE Trans. on Multimedia*, to appear.

84. K. L. Ginter, V. H. Shear, F. J. Spahn, and D. M. Van Wie. System and methods for secure transaction management and electronic rights proctection. In *United States Patent 5,917,912*, 1999.

85. M. Ramkumar and A. N. Akansu. A capacity estimate for data hiding in internet multimedia. In *Content Security and Data Hiding in Digital Media*, 1999.

86. C. H. Chou and Y. C. Li. A perceptually tuned subband image coder based on the measure of just-noticeable-distortion profile. *IEEE Trans. on Circuits, Systems and Video Technology*, 5(6):467–476, 1995.

87. G. E. Legge and J. M. Foley. Contrast masking in human vision. *Journal of the Optical Society of America*, 70(12):1458–1471, 1980.

88. M. D. Swanson, B. Zhu, and A.H. Tewfik. Transparent robust image watermarking. In *IEEE International Conference on Image Processing*, volume 3, pages 211–214, 1996.

89. D. Coltuc and P. Bolon. Watermarking by histogram specification. In *SPIE Security and Watermarking of Multimedia Contents*, volume 3657, pages 252–263, 1999.

90. P. M. Rongen, M. J. Maes, and K. W. van Overveld. Digital image watermarking by salient point modification: Practical results. In *SPIE, Security and Watermarking of Multimedia Contents*, volume 3657, pages 273–282, 1999.

91. A. V. Oppenheim and J. S. Lim. The importance of phase in signals. *Proc. of the IEEE*, 69:529–541, 1981.

92. W. A. Pearlman and R. M. Gray. Source coding of the discrete fourier transform. *IEEE Transactions on Information Theory*, IT-24:683–692, 1978.

93. A.G. Tescher. The role of phase in adaptive image coding. In *Tech USCIPI Pub. 510, Image Processing Institute,* University of Southern California, 1978.

94. H. T. Sencar, M. Ramkumar, and A. N. Akansu. Improvements on data hiding for lossy compression. In *IEEE International Conference on Acoustics, Speech and Signal Processing*, volume 2, pages 444–447, 2002.

95. G. K. Wallace. The jpeg still picture compression standard. *Communications of the ACM*, 34(4):31–44, April 1991.

96. W. Zeng and B. Liu. On resolving rightful ownerships of digital images by invisible watermarks. In *IEEE International Conference on Image Processing*, volume 1, pages 552–555, 1997.

97. N. Memon and P. W. Wong. A buyer-seller watermarking protocol. In *IEEE Second Workshop on Multimedia Signal Processing*, pages 273–278, 1998.

98. M. Ramkumar and A. N. Akansu. A robust scheme for oblivious detection of watermarks/data hiding in still images. In *Proc. of SPIE Multimedia Systems and Applications*, volume 3528, pages 474–481, 1998.

99. R. B. Wolfgang and E. J. Delp. Overview of image security techniques with applications in multimedia systems. In *SPIE Conference on Multimedia Networks: Security, Displays, Terminals, and Gateways*, volume 3228, pages 297–308, 1997.

100. B. Schneier. *Applied Cryptography, Second Edition*. New York: Wiley & Sons, 1996.

101. A. Papoulis. *Probability, Random Variables, and Stochastic Processes, Third Edition*. Singapore: McGraw Hill Inc, 1991.

102. M. Ramkumar and A. N. Akansu. On the design of robust data hiding systems. In *Proc. of 33rd ASILOMAR Conference on Signals, Systems and Computers*, 1999.

103. M. Ramkumar and A. N. Akansu. Optimal design of data hiding methods robust to lossy compression. *IEEE Trans. on Multimedia*, to appear.

104. I. B. Ozer, M. Ramkumar, and A. N. Akansu. A new method for detection of watermarks in geometrically distorted images. In *IEEE International Conference on Acoustics, Speech and Signal Processing*, volume 4, pages 1963–1966, 2000.

105. J. J. K. Ó. Ruanaidh and T. Pun. Rotation, scale and translation invariant spread spectrum digital image watermarking. *Signal Processing*, 66(3):303–317, 1998.

106. L. Qiao and K. Nahrstedt. Watermarking schemes and protocols for protecting rightful ownership and customer's rights. *Journal of Visual Communication and Image Representation*, 9:194–210, 1998.

107. M. Ramkumar and A. N. Akansu. Self-noise suppression schemes for blind image steganography. In *Proc. of SPIE Multimedia Systems and Applications (Image Security)*, volume 3845, pages 474–481, 1999.

108. M. Ramkumar and A. N. Akansu. A robust oblivious watermarking scheme. In *IEEE International Conference on Image Processing*, volume 4, pages 1231–1235, September 2000.

109. R. A. Fisher. *Statistical Methods for Research Workers*. New York: Hafner Press, 1970.

110. M. Wu and B. Liu. *Multimedia Data Hiding*. New York: Springer Verlag, 2003.

Index